# MATLAB® Essentials
## A First Course for Engineers and Scientists

# MATLAB® Essentials
## A First Course for Engineers and Scientists

William Bober

CRC Press
Taylor & Francis Group
Boca Raton  London  New York

CRC Press is an imprint of the
Taylor & Francis Group, an **informa** business

MATLAB® is a trademark of The MathWorks, Inc. and is used with permission. The MathWorks does not warrant the accuracy of the text or exercises in this book. This book's use or discussion of MATLAB® software or related products does not constitute endorsement or sponsorship by The MathWorks of a particular pedagogical approach or particular use of the MATLAB® software.

CRC Press
Taylor & Francis Group
6000 Broken Sound Parkway NW, Suite 300
Boca Raton, FL 33487-2742

© 2018 by Taylor & Francis Group, LLC
CRC Press is an imprint of Taylor & Francis Group, an Informa business

No claim to original U.S. Government works

Printed on acid-free paper

International Standard Book Number-13: 978-1-138-56328-5 (Hardback)
International Standard Book Number-13: 978-1-138-03237-8 (Paperback)

This book contains information obtained from authentic and highly regarded sources. Reasonable efforts have been made to publish reliable data and information, but the author and publisher cannot assume responsibility for the validity of all materials or the consequences of their use. The authors and publishers have attempted to trace the copyright holders of all material reproduced in this publication and apologize to copyright holders if permission to publish in this form has not been obtained. If any copyright material has not been acknowledged please write and let us know so we may rectify in any future reprint.

Except as permitted under U.S. Copyright Law, no part of this book may be reprinted, reproduced, transmitted, or utilized in any form by any electronic, mechanical, or other means, now known or hereafter invented, including photocopying, microfilming, and recording, or in any information storage or retrieval system, without written permission from the publishers.

For permission to photocopy or use material electronically from this work, please access www.copyright. com (http://www.copyright.com/) or contact the Copyright Clearance Center, Inc. (CCC), 222 Rosewood Drive, Danvers, MA 01923, 978-750-8400. CCC is a not-for-profit organization that provides licenses and registration for a variety of users. For organizations that have been granted a photocopy license by the CCC, a separate system of payment has been arranged.

**Trademark Notice:** Product or corporate names may be trademarks or registered trademarks, and are used only for identification and explanation without intent to infringe.

**Visit the Taylor & Francis Web site at**
**http://www.taylorandfrancis.com**

**and the CRC Press Web site at**
**http://www.crcpress.com**

Printed and bound in the United States of America by Sheridan

# Contents

Preface ..................................................................................................................ix
Acknowledgments ...........................................................................................xiii
Author ................................................................................................................xv

1. **Computer Programming with MATLAB® for Engineers** ....................1
   1.1 Introduction ............................................................................................1
   1.2 Computer Usage in Engineering ..........................................................1
   1.3 Mathematical Model ..............................................................................2
   1.4 Computer Programming .......................................................................3
   1.5 Components of a Typical Desktop/Laptop Computer System .....3
   1.6 Overview of Programming Languages ...............................................5
   1.7 Why MATLAB? .......................................................................................5
   1.8 Programming Methodologies ...............................................................6
   1.9 MATLAB Programming Language .....................................................6
   1.10 Building Blocks in Writing a Computer Program ............................7
   1.11 Example Programs ................................................................................7

2. **MATLAB® Fundamentals** ........................................................................9
   2.1 Introduction ............................................................................................9
   2.2 MATLAB's Desktop ............................................................................10
   2.3 Constructing a Script (Program) in MATLAB ................................13
   2.4 Variable Names and Types .................................................................18
   2.5 Assignment Operator ..........................................................................19
       2.5.1 Arithmetic Operators ...............................................................20
   2.6 Some MATLAB Features, Commands, Special Items,
       and Built-in Functions ........................................................................21
       2.6.1 Trigonometric and Other Useful Functions .......................21
             2.6.1.1 Special Values ..........................................................23
             2.6.1.2 Trigonometric Functions ......................................24
             2.6.1.3 Inverse Trigonometric Functions ........................24
             2.6.1.4 Exponential, Square Root, and Error
                     Functions ..................................................................25
             2.6.1.5 Complex Numbers .................................................26
       2.6.2 Other Special Values ................................................................26
             2.6.2.1 Other Useful MATLAB Functions .....................26
             2.6.2.2 Colon Operator (:) ..................................................28
             2.6.2.3 Preallocation of a Matrix .......................................29

| | | | |
|---|---|---|---|
| 2.7 | | MATLAB Output | 30 |
| | 2.7.1 | The `disp` Command | 31 |
| | 2.7.2 | The `fprintf` Command | 31 |
| | 2.7.3 | Printing to a File | 32 |
| 2.8 | | Simple Plot Commands | 34 |
| | 2.8.1 | Linear Plot | 34 |
| 2.9 | | Loops | 36 |
| | 2.9.1 | The `for` Loop | 36 |
| | 2.9.2 | The `While` Loop | 43 |
| 2.10 | | Input | 50 |
| | 2.10.1 | The `Load` Command | 50 |
| | 2.10.2 | The `dlmread` Command | 52 |
| | 2.10.3 | `fscanf` Command | 52 |
| | 2.10.4 | The `input` Command | 54 |
| 2.11 | | More on MATLAB Graphics | 55 |
| | 2.11.1 | The `figure` Command | 55 |
| | 2.11.2 | Multiple Plots | 57 |
| | 2.11.3 | The `hold on` Command | 59 |
| | 2.11.4 | `Plotyy` Command | 62 |
| | 2.11.5 | The `subplot` Command | 63 |
| | 2.11.6 | Bar Charts | 63 |
| | 2.11.7 | Pie Charts | 65 |
| | 2.11.8 | Greek Letters and Mathematical Symbols | 68 |
| | 2.11.9 | Interactively Annotating Plots | 69 |
| | 2.11.10 | Saving Plots | 69 |
| References | | | 80 |

## 3. Conditional Operators, Built-in Functions with Vector Arguments, MATLAB®'s Interp1 Function, and Some Scalar and Vector Operations ............81

| | | | |
|---|---|---|---|
| 3.1 | | Introduction | 81 |
| 3.2 | | Conditional Operators and Alternate Paths | 81 |
| | 3.2.1 | The `if` Command Provides Two Alternate Paths | 81 |
| | 3.2.2 | The `if-elseif-else` Command Provides More than Two Alternate Paths | 83 |
| | 3.2.3 | The `break` Command | 85 |
| | 3.2.4 | The `switch` Command | 89 |
| | 3.2.5 | MATLAB's `menu` Function | 90 |
| 3.3 | | Working with Built-in Functions with Vector Arguments | 92 |
| 3.4 | | MATLAB's `interp1` Function | 93 |
| 3.5 | | Some Scalar and Vector Operations | 96 |
| | 3.5.1 | Addition of a Scalar and a Vector | 96 |
| | 3.5.2 | Multiplication of a Scalar Times a Vector | 96 |
| | 3.5.3 | Addition and Subtraction of Two Vectors of the Same Length | 96 |

Contents      vii

    3.5.4    Element-by-Element Operations ................................... 96
    3.5.5    Operation of Two Vector Functions ............................ 98

## 4. Self-Written Functions and MATLAB®'s `fminbnd` Function ........... 105
    4.1    Introduction ................................................................................. 105
    4.2    Self-Written Function .................................................................. 105
    4.3    Anonymous Functions ................................................................ 110
    4.4    MATLAB's `fminbnd` ................................................................. 113
    References ............................................................................................ 122

## 5. Working with Characters and Strings ................................................. 123
    5.1    Introduction ................................................................................. 123
    5.2    MATLAB's `textscan` Function ................................................ 127

## 6. Roots of Algebraic and Transcendental Equations ............................ 131
    6.1    Introduction ................................................................................. 131
    6.2    Search Method ............................................................................. 132
    6.3    Bisection Method ........................................................................ 133
    6.4    MATLAB's `fzero` Function ..................................................... 134
    6.5    MATLAB's `roots` Function ..................................................... 139
    References ............................................................................................ 152

## 7. System of Algebraic, Linear Equations ............................................... 153
    7.1    Introduction ................................................................................. 153
    7.2    System of Algebraic, Linear Equations ..................................... 153
        7.2.1    MATLAB's `inv` Function ............................................ 154
        7.2.2    Gauss-Elimination Method ........................................... 154
    7.3    Treatment of Large Systems of Algebraic, Linear Equations ...... 156
    7.4    A Resistive Circuit Problem ....................................................... 159
    7.5    Gauss Elimination ....................................................................... 161
    7.6    Number of Solutions ................................................................... 162
    References ............................................................................................ 167

## 8. Curve Fitting ............................................................................................ 169
    8.1    Introduction ................................................................................. 169
    8.2    MATLAB's Curve-Fitting Functions ......................................... 169
    8.3    Curve Fitting with the Exponential Function ......................... 174
    8.4    Cubic Splines ............................................................................... 178
        8.4.1    MATLAB's Cubic Spline Curve-Fitting Function ........... 179

## 9. Numerical Integration ........................................................................... 187
    9.1    Introduction ................................................................................. 187
    9.2    Numerical Integration and Simpson's Rule ............................ 187
    9.3    Improper Integrals ...................................................................... 190

|  |  |  |
|---|---|---|
| | 9.4 | MATLAB's `integral` Function ................................................. 190 |
| | 9.5 | MATLAB's `integral2` Function ............................................... 194 |
| | Reference ............................................................................................. 204 |

**10. Numerical Integration of Ordinary Differential Equations ............. 205**
    10.1    Introduction ........................................................................... 205
    10.2    Initial Value Problem and MATLAB's `Ordinary Differential Equations` Function ........................................ 206
    Reference ............................................................................................. 226

**11. Boundary Value Problems of Ordinary Differential Equations ...... 227**
    11.1    Introduction ........................................................................... 227
    11.2    Difference Formulas ............................................................. 227
    References ........................................................................................... 235

**Appendix: Greek Letters and Special Characters in MATLAB® Plots ..... 237**

**Review Answers ........................................................................................ 241**

**Index ........................................................................................................... 255**

# *Preface*

I have taught computer applications course for engineers in the mechanical and civil engineering departments at Florida Atlantic University (FAU), Boca Raton, Florida, for many years. I first started teaching the course using the Fortran language. Some years later, the department switched to the C/C++ language. More recently, the course has been taught using MATLAB®. The advantage of using MATLAB over many other programming languages is that MATLAB contains functions that enable the user to solve various mathematical problems, such as interpolation, roots of algebraic equations, the relative minimum and maximum of a function, a system of linear algebraic equations, curve-fitting problems, definite integrals, a system of ordinary differential equations, and many others, some of which require special tool boxes at an extra cost. There are also programming techniques available in MATLAB but not available in either Fortran or C/C++. Because not all engineering firms use MATLAB, I decided, in writing this textbook, to first cover some very basic building blocks applicable to most, if not all, computer programming languages used by engineers before getting into programming that is specific to MATLAB. The syntax of these basic building blocks may be different in different languages, but the concept is the same. The basic building blocks in programming covered in Chapters 2 through 4 are as follows:

1. Variable types, scalars, vectors, and matrices
2. Assignments (which, in most cases in this book, are arithmetic statements)
3. Input/output statements
4. Loop statements
5. Conditional operators
6. Functions (built-in and self-written)

Before MATLAB, it was rare that I would write a computer program without using a `for` loop. With that in mind, in this textbook, I introduce `for` loops as early as is feasible. The authors of most other MATLAB textbooks introduce `for` loops at a much later stage in their books.

Although students at FAU take the computer programming course in their sophomore year, having taken Calculus II, the textbook can also be used at the freshman level (the first eight chapters do not involve calculus). Although there are many engineering example applications, the governing equations are given without derivations. Therefore, students not only see variables $x$ and $y$ but also see variables of pressure ($p$), temperature ($T$), time ($t$), velocity (V), voltage (v), current (i), and so on. The chapters include review sections, which may be used by the course instructor to ask the class questions on the material that has been recently covered.

The primary objectives of the textbook are as follows:

1. To teach the reader the basic concepts in writing a computer program (script) on the MATLAB platform, although many of the concepts taught are also applicable to other computer programming languages.
2. To familiarize the reader with many of MATLAB's built-in functions, some of which can be used to solve several mathematical problems, such as interpolating for properties between table values, finding the roots of transcendental and polynomial equations, determining the relative minimum or maximum of a function, and solving a system of linear algebraic equations and curve fitting. The last two chapters involve calculus and thus would only be applicable for a course at the sophomore or higher level. These last two chapters cover MATLAB's functions for determining the value of a definite integral and for solving a system of ordinary differential equations.

I have tried to organize the material so that the student gets to write a meaningful program within several weeks of starting the course. The students are required to add a comment section to their programs describing what the program is about. Nearly all exercises and projects require the student to produce tables or graphs or both.

The text contains many complete sample MATLAB programs and their results, including tables, graphs, and comments what the program is about. These examples should provide guidance to the student on completing the exercises and projects that are listed in each chapter. Projects are at the end of the chapters and are usually more difficult than the exercises. Many of the projects are nontrivial. In recent times, I have used several exercise problems as in-class exams in which students submit their MATLAB programs and results to me on blackboard. Projects are given as take-home exams to be submitted to me within 1 or 2 weeks, depending on the difficulty of the project. The projects require the student to write a computer program in MATLAB to solve a mathematical or engineering-type problem.

The computer applications course that I teach is run as a lecture-laboratory course. The advantage of running the course in this manner is that the instructor is in the computer laboratory to help the student debug his or her program. This includes the example programs as well as the exercises and the projects. See the Table of Contents to get a more complete description of the material covered in this textbook.

*Preface*  xi

All example scripts in this book are available for download on the CRC Press Website at https://www.crcpress.com/MATLAB-Essentials-A-First-Course-for-Engineers-and-Scientists/Bober/p/book/9781138032378.

MATLAB® is a registered trademark of The MathWorks, Inc. For product information, please contact:

The MathWorks, Inc.
3 Apple Hill Drive
Natick, MA 01760-2098 USA
Tel: 508-647-7000
Fax: 508-647-7001
E-mail: info@mathworks.com
Web: www.mathworks.com

# *Acknowledgments*

I thank Jonathan Plant of CRC Press for his confidence and encouragement in writing this textbook. I thank Dr. Andrew Stevens for allowing me to extract many electrical engineering concepts and projects from our joint textbook titled *Numerical and Analytical Methods with MATLAB for Electrical Engineers*. I also thank Ed Curtis and Bala Gowri for guiding me through the textbook submission process. I also thank the following people for their graphic contributions: Danielle Mitchell and Jacqueline Ferrer. Finally, I wish to express my deep gratitude to my wife for tolerating the many hours I spent on preparation of this manuscript—time which otherwise would have been devoted to my family.

# *Author*

**William Bober**, PhD, earned a BS in civil engineering at the City College of New York (CCNY), New York, New York, an MS in engineering science at Pratt Institute, Brooklyn, New York, and a PhD in engineering science and aerospace engineering at Purdue University, West Lafayette, Indiana. At Purdue University, he was on a Ford Foundation Fellowship and was assigned to teach one engineering course in each semester. After he completed his doctoral work, he worked as an associate engineering physicist in the Applied Mechanics Department at Cornell Aeronautical Laboratory, Buffalo, New York. After leaving Cornell Labs, he was employed as an associate professor in the Department of Mechanical Engineering at the Rochester Institute of Technology (RIT), Rochester, New York, for 12 years. After leaving RIT, he accepted a position as an associate professor in the Department of Mechanical Engineering at the Florida Atlantic University (FAU). Recently, he was transferred to the Department of Civil Engineering at FAU. He taught a computer application course for engineers for many years at both RIT and FAU. This experience has given him the knowledge to write or coauthor several textbooks on the subject, including *Numerical and Analytical Methods with MATLAB, Numerical and Analytical Methods with MATLAB for Electrical Engineers,* and *Introduction to Analytical Methods with MATLAB for Engineers and Scientists*, all published by CRC Press. This textbook, *MATLAB Essentials: A First Course for Engineers and Scientists*, is for a first course in computer applications on the MATLAB platform for engineers and scientists. He has also written several papers for the *International Journal of Mechanical Engineering Education* (IJMEE).

# 1

# Computer Programming with MATLAB® for Engineers

## 1.1 Introduction

Most, if not all, engineering companies use computers in one way or the other. Many employ computer programmers to solve company-specific problems. These companies may also purchase or license software packages such as C/C++ or MATLAB®, and install the programs on their computer systems to enable their programmers and engineers to efficiently solve company-specific problems. The field of engineering, in particular, lends itself to analytical and numerical solutions due to the highly mathematical nature of the field. Analytical and numerical methods invariably involve writing computer code to solve a problem of interest. Mathematical methods for solving many types of engineering problems use concepts from linear algebra, root extraction of polynomial and transcendental equations, integration, curve fitting, differential equations, and so on. MATLAB, using a variety of analytical and numerical methods, has created built-in functions that enable the user to readily employ these mathematical methods. However, the user needs to know some programming techniques to effectively make use of these built-in functions. Many examples involving the use of these built-in functions are covered in this textbook.

## 1.2 Computer Usage in Engineering

Some of the ways that the computer is used in engineering are as follows:

1. Solving mathematical models of physical phenomena
2. Storing and reducing experimental data
3. Controlling machine operations
4. Communicating with other engineers and technicians on a particular project

This textbook is mostly concerned with item 1, that is, solving mathematical models of physical phenomena.

The engineer's interest lies in

1. Designing new products or improving existing ones
2. Improving manufacturing efficiency
3. Minimizing cost and power consumption
4. Maximizing yield and return on investment
5. Minimizing time to market
6. Research on developing new products

These can be accomplished by

1. Full-scale experiments. May be prohibitively expensive.
2. Small-scale model experiments. Still very expensive, and extrapolation is frequently questionable.
3. A mathematical model that is the least expensive and faster. It can provide more detailed answers and different cases under different conditions and can be run quickly. If there is confidence in a mathematical model, it will be used in preference to experiment.

## 1.3 Mathematical Model

Physical phenomena are described by a set of governing equations. Numerical methods are frequently used to solve the set of governing equations, since closed-form solutions for many types of problems are not available. Even when closed-form solutions are available, the solution may be sufficiently complicated that the computer is needed to calculate the desired answer. Numerical methods invariably involve the computer. The computer performs arithmetic operations upon discrete numbers in a defined sequence of steps. The sequence of steps is defined in the program. A useful solution is obtained if

1. The mathematical model accurately represents the physical phenomena; that is, the model has the correct governing equations.
2. The numerical method is accurate.

    Note: If the governing equations are not correct, the solution will be worthless regardless of the accuracy.

3. The numerical method is programmed correctly.
4. This book is mainly concerned with items (2) and (3).

## 1.4 Computer Programming

The advantage of using the computer is that it can carry out many calculations in a fraction of a second; at the time of this writing, computer speeds are measured in teraflops (trillions of floating-point operations per second). However, to leverage this power, we need to write a set of instructions, that is, a program or script. For the problems of interest in this book, the digital computer is only capable of performing arithmetic, logical, and graphical operations. Therefore, arithmetic procedures must be developed for evaluating integrals, determining roots of a transcendental equation, solving a system of linear equations, solving differential equations, and so on. The arithmetic procedure usually involves a set of algebraic equations. A computer solution for such problems involves developing a computer program that defines a step-by-step procedure for obtaining an answer to the problem of interest. The method of solution is called an *algorithm*. Depending on the particular problem, we might write our own algorithm, or, as we shall see, we can also use the algorithms built into a package like MATLAB in order to carry out well-known algorithms for solving many types of mathematical problems.

## 1.5 Components of a Typical Desktop/Laptop Computer System

1. Input devices, typically includes a keyboard and mouse, but may also include a touch-screen, a microphone, or a similar device. Input devices provide a mechanism for humans to provide data and instructions to the computer.
2. A Central Processing Unit (CPU) consisting of a Control Unit, an Arithmetic Logic Unit (ALU), and registers. The Control Unit fetches instructions from memory, executes the instructions, and then returns the results to memory. The ALU performs arithmetic and logical operations. Registers are high-speed local memory locations, and are used to provide operands and store results from the ALU.
3. A Memory/Storage Unit in which data and instructions are stored. There are two types of memory: Main Memory and Secondary Memory. Main Memory is used for temporary storage of programs and data, and is commonly implemented with Dynamic Random Access Memory (DRAM) devices. Items in Main Memory are not saved when the computer is shut off.

4. Secondary Memory stores data permanently. The Secondary Memory commonly consists of
   a. *Hard drive*: Provides semipermanent storage of programs and data. It is usually internal to the computer. It has a large storage capacity (terabytes).
   b. *Optical drive (DVD)*: Stores programs, data, video onto a CD, or a DVD disk for permanent storage.
   c. *Flash drive*: Stores programs and data onto a removable flash memory stick, which can be used to transfer programs and data from one computer to another.
5. Output devices, typically include a monitor or a printer, but may also include speakers, a projector, a VR headset, or a similar device. Output devices provide a mechanism for humans to receive data, sound, and images from the computer.
6. Network interface allows the computer to send/receive data to other computers in the vicinity (a local area network, or LAN) or around the workplace (a wide area network, or WAN). Typical applications that utilize the network include file transfer, e-mail, World Wide Web, and streaming audio/video.
7. The operating system (OS) provides a unified environment for software to utilize and control the computer. The OS manages storage in order to enable the creation of *files* that are organized in *folders* and are stored to a *C drive*. The OS also controls *booting* the computer and instructs the display to show the *desktop*, and receives signals from the mouse in order to move the cursor. Common operating systems include Windows, MacOS, and Linux.

MATLAB utilizes all of the computer components described above. Users utilize a keyboard and mouse to write and execute scripts. The OS saves the scripts to storage and loads the MATLAB executable program into the memory. MATLAB specifies instructions for execution by the CPU, and subsequent display of results to the monitor or printer. The network allows download of documentation, access to remote MATLAB servers, and software updates.

The memory of a computer is an ordered sequence of storage locations called memory cells. Each memory cell has an address indicating its relative position in memory. The memory cell is a collection of smaller units called bytes. A byte is the amount of storage required to store a single character (letter, number, or symbol). A byte is a collection of smaller units called bits. A bit takes on the value of 0 or 1, and is therefore suited for the binary system of numbers. Generally, there are eight bits to a byte. Each character or set of characters, or value, is represented by a particular pattern of zeros and ones. The computer can retrieve or store a value.

## 1.6 Overview of Programming Languages

A program is a list of instructions to be carried out by the computer. There are two types of software:

1. *System Software*: Performs tasks required for the operation of the computer, such as Windows 10, Windows 7, Unix, Linux, and so on.
2. *Application Software*: Written to perform particular tasks for the person using the computer. These would include programs such as Microsoft's Office, AutoCAD, MATLAB, and so on. Programs written by individuals would also be classified as Application Software.

There are different levels of computer languages. However, the computer can only execute programs in machine language, which is considered the lowest level. All higher level language instructions must be translated into machine language. A sequence of machine language consists of a collection of zeros and ones. Higher level languages include the following:

MATLAB, FORTAN, Basic, C/C++, Pascal, and so on. These higher level languages allow as to write programs in a more familiar and understandable manner than a program in machine language.

## 1.7 Why MATLAB?

MATLAB was originally written by Dr. Cleve Moler at the University of New Mexico, Albuquerque, NM in the 1970s and was commercialized by MathWorks, Natick, MA in the 1980s. It is a general purpose numerical package that allows complex equations to be solved efficiently, and subsequently generate tabular or graphical output. Although there are many numerical packages available to engineers, many are very highly focused toward a particular application, for example, ANSYS for modeling structural problems using the finite element method. As of the time of this writing, MATLAB R2016b runs natively on Microsoft Windows, Apple Mac OS, and Linux. In this textbook, we will assume that you are running MATLAB on your local machine in a Microsoft Windows environment. It should be straightforward for non-Windows users to translate the usage descriptions to their preferred environment. In any case, these differences are largely limited to the cosmetics and presentation of the program, and not the MATLAB commands themselves. All versions of MATLAB (on any platform) use the same command set, and the Command Window on all platforms should behave identically.

MATLAB is offered with accompanying *toolboxes* at additional cost to the user. A wide variety of toolboxes are available in fields such as statistics, optimization, control systems, and so on. However, in this textbook, we will concentrate on teaching basic elements in computer coding on the MATLAB platform and on fundamental numerical concepts without requiring the purchase of any additional toolboxes.

## 1.8 Programming Methodologies

There are many methodologies for computer programming, but the tasks at hand boil down to

1. Studying the problem to be programmed including the geometry of the problem.
2. Listing the algebraic equations specified in the problem statement. The equations will be based on the known physical phenomena.
3. Selecting the most efficient computer code and numerical method for obtaining a solution to the problem of interest.
4. Creating an outline or a flow chart for the program flow (today, not many textbooks on MATLAB recommend creating a flow chart).
5. Writing the program using the list of algebraic equations and the outline or flow chart.
6. Debugging the program by running it and fixing any syntax errors (programming language errors).
7. Examining the solution to see if it makes sense.
8. Refining and further debugging the algorithm and program flow.

Experienced programmers often omit some of these steps (or do them in their head), but the overall process resembles any engineering project: design, create a prototype, test, and iterate the process until a satisfactory product is achieved.

## 1.9 MATLAB Programming Language

MATLAB may be considered a programming or scripting language unto itself, but like every programming language, it has the following core components:

1. Data types, that is, integers, floating-point numbers, strings, vectors, and matrices.

2. Operators and built-in functions (e.g., commands for addition, subtraction, multiplication, division, trigonometric functions, and log function).
3. Control flow directives for making decisions and performing repeated operations (e.g., loops, alternate paths, and functions).
4. Input/output ("I/O") commands for receiving input from a user or a file and for generating output to a file or to the screen (e.g., read and print statements).

MATLAB borrows many constructs from other languages. For example, the `for` and `while` loops and the `fprintf` commands are from the C programming language (or its descendents C++). However, the biggest difference is that the basic element in MATLAB is a matrix, thus providing the ability to manipulate large amounts of data with a terse syntax, and allowing for the solution of complicated problems in just a few lines of code. In addition, MATLAB is also very rich in presentation functions to display sophisticated plots and graphs.

## 1.10 Building Blocks in Writing a Computer Program

Most engineering computer programs will include some or all of the following building blocks in program development:

1. Variable types, scalars, vectors, and matrices
2. Assignments (which, in most cases in this book, is an arithmetic statement)
3. Input/Output statements
4. Loop statements
5. Conditional Operators
6. Functions (built-in and self-written)

Example programs containing these program building blocks are given throughout this book.

## 1.11 Example Programs

The example programs in this book may be downloaded from the publisher's website at https://www.crcpress.com/MATLAB-Essentials-A-First-Course-for-Engineers-and-Scientists/Bober/p/book/9781138032378. Students may then run the example programs on their own computer and see the results. It may also be beneficial for students to type-in a few of the sample programs (along with some inevitable syntax and typographical errors), thereby giving the student the opportunity to see how MATLAB responds to program errors and subsequently learn what they need to do to fix the problem.

### REVIEW 1.1

1. List several ways engineers use the computer.
2. List several areas of interest for engineers.
3. List several methods that can be used in the design of new products.
4. Which method mentioned in item 3 is the least expensive?
5. List several components of a typical desktop/laptop computer system.
6. Name several computer languages used today and in the past by Engineers.
7. What is the lowest level computer language and what numbering system does it use?
8. For engineers, what is the principle advantage of using MATLAB over several of the other computer programming languages?
9. List several recommendations in developing a computer program for solving a particular problem.
10. List several building blocks available in developing a program in MATLAB or in most other Engineering Software Platforms such as C or C++.

# 2
# MATLAB® Fundamentals

## 2.1 Introduction

MATLAB® is a software program for numeric computation, data analysis, and graphics. One advantage that MATLAB has for engineers over programming languages such as C or C++ is that the MATLAB program includes functions that numerically solve

1. Large systems of linear algebraic equations.
2. Roots of transcendental and polynomial equations.
3. One- and two-dimensional definite Integrals.
4. A system of first-order ordinary differential equations.
5. Statistical problems.
6. Optimization problems.
7. Control systems problems.
8. Many other types of problems encountered in engineering.

MATLAB also offers toolboxes (which must be purchased separately) that are designed to solve problems in specialized areas.

In this chapter, we first familiarize the reader with some of the basic elements of the MATLAB platform. This allows the reader to shortly learn to do computations in the Command Window and to write and run a script in MATLAB. This is followed by discussing the basic building blocks in constructing a computer program (script) for solving mathematical- and engineering-type problems on the MATLAB platform. These building blocks are applicable in any programming language, the syntax may be different, but the concept is the same. Recall at the end of Chapter 1, the building blocks that was mentioned and which are covered in this book are

1. Variable types, scalars, vectors, and matrices.
2. Assignments (including arithmetic statements).
3. Input/Output statements.

4. Loop statements.
   5. Conditional Operators (leading to alternate paths in the program).
   6. Functions (built-in and self-written).

Items (1)–(4) are covered in this chapter as well as some of the elementary built-in functions of item (6). Items (5) and (6) are covered in Chapter 3. Examples of MATLAB programs that solve various types of mathematical problems, many of which are related to engineering-type problems, are covered throughout this book.

## 2.2 MATLAB's Desktop

Mathworks, the company that developed MATLAB, normally update their version of MATLAB every six months. In this textbook, Sections 2.2 and 2.3 discusses the MATLAB desktop windows and how to construct a script in MATLAB based on MATLAB version R2016b.

Under Microsoft Windows, MATLAB may be started via the Start Menu or by clicking on the MATLAB icon on the desktop. Upon startup, a new window will open containing the MATLAB *desktop* (not to be confused with the Windows desktop), and one or more MATLAB windows will open within the MATLAB desktop (see Figure 2.1 for the default configuration).

The main windows are the Command Window, Current Folder, and Workspace. You can customize the MATLAB windows that appear upon startup by clicking on *Layout* in the Toolstrip and checking (or unchecking) the windows that you wish to appear on the MATLAB desktop. Figure 2.1 shows the Command Window (in the center), the Current Folder Window (on the left), the Workspace Window (on the right), and a Long Narrow box containing the Path to the Current Folder (just below the Toolstrip and just above the Command Window). MATLAB designates this Long Narrow box as the Current Folder Toolbar. These windows and the Current Folder Toolbar are summarized as follows:

- *Command Window*: In the Command Window, you can enter commands and data, make calculations, and print results. You can write a script in the Command Window and execute the script. However, writing a script directly into the Command Window is discouraged because it will not be saved, and if an error is made,

# MATLAB® Fundamentals    11

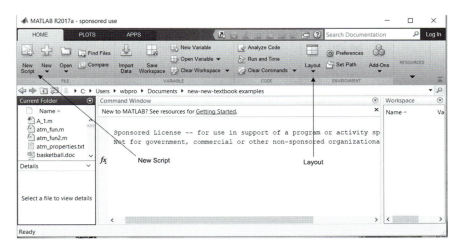

**FIGURE 2.1**
MATLAB desktop windows for the three-column option.

the entire script must be retyped. By using the up arrow (↑) key on your keyboard, the previous command can be retrieved (and edited) for reexecution.

- *Current folder toolbar*: This Toolbar gives the path to the Current Folder. *To run a MATLAB script, the script needs to be in the folder listed in this Toolbar.*
- *Current Folder Window (on the left)*: This window lists all the files in the Current Folder whose path is listed in the Current Folder Toolbar. By double clicking on a file in this window, the file will open within MATLAB.
- *Workspace Window*: This window will be on the right for a three-column option (see Figure 2.1) or below the Current Folder Window for the two-column option (see Figure 2.2). The two- or three-column option can be selected from the *layout* options in the Toolstrip. The Workspace Window contains all the commands entered into the Command Window.
- *Editor Window*: To open this window, click on the *New Script icon* in the Toolstrip in MATLAB's desktop (see Figure 2.1). This will open the Editor Window (see Figure 2.3). This window may be used to create, edit, and execute MATLAB scripts (also called programs). Figure 2.4 contains the script for Example 2.1.

12                                                                                        MATLAB® Essentials

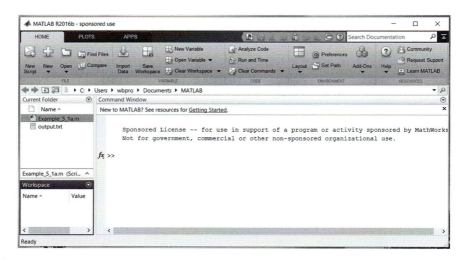

**FIGURE 2.2**
MATLAB desktop windows for two-column option.

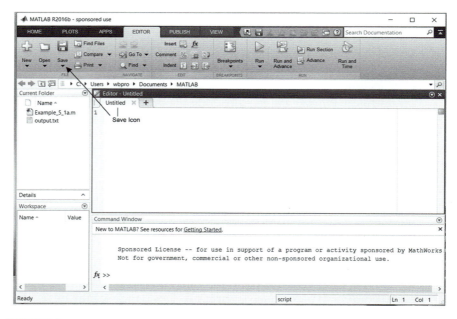

**FIGURE 2.3**
Editor Window just above the Command Window.

# MATLAB® Fundamentals

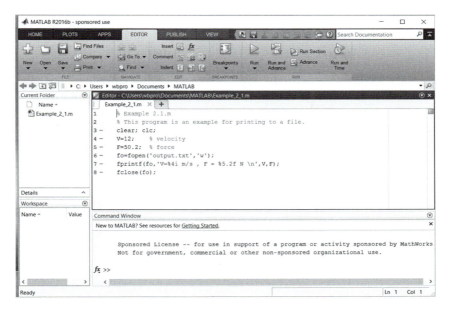

**FIGURE 2.4**
Script for Example 2.1.

## 2.3  Constructing a Script (Program) in MATLAB

In the first few examples the reader is asked to type-in several commands in the Command Window. Subsequent examples involve creating scripts in the Editor Window. The following list summarizes the steps for writing a script in MATLAB:

1. If available, start the MATLAB program by double-clicking on the MATLAB icon on the Window's desktop. If not available, go to the Window's *Start* Menu, click on *All Programs*, find the MATLAB program among the list of available programs, and double-click on it. This will open up the MATLAB desktop.
2. Click on the *New Script icon* in the Toolstrip in MATLAB's desktop. This brings up a new *Editor Window* just above the Command Window (see Figure 2.3).
3. Type your program into the Editor Window.

4. When you are finished typing in the program, save the script by clicking on the *Save* icon in the Toolstrip (see Figure 2.3). A dialog box will open in which you are to select the folder (left column), and in which you are to type-in the name of the script in the File Name Dialog Box (see Figure 2.5). By default, your program will be saved with the *.m* extension. It is best to use a folder that contains **only** your own MATLAB scripts.

5. You may then run the script in the Editor Window by clicking on the arrow located just above the *Run icon* in the Toolstrip (see Figure 2.6). This icon is a *Save and Run* Command. Note: In the Editor Window, the arrow is green.

Alternatively, you can run the script from the Command Window by typing the script name (without the *.m* extension) after the MATLAB prompt (>>). For example, if the program has been saved as *heat.m*, then type *heat* after the MATLAB prompt (>>), as shown below:

>> heat

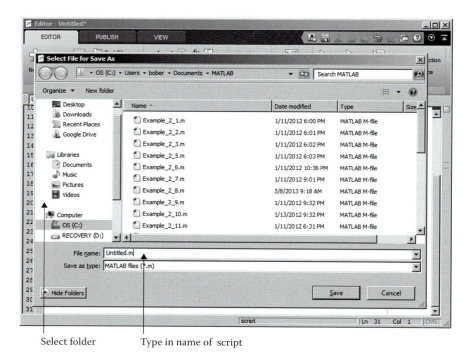

**FIGURE 2.5**
Select folder (left column) and type in the name of the script in the file name dialog box.

# MATLAB® Fundamentals 15

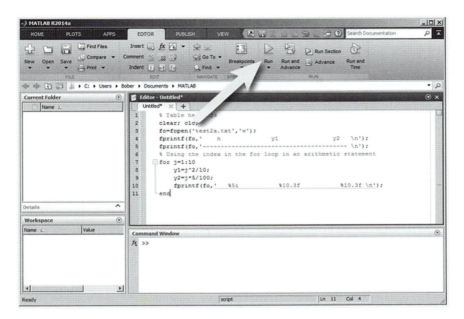

**FIGURE 2.6**
Save and Run icon in the toolstrip.

6. If you try to run your script and your script is not in the Current Folder whose path is listed in the Current Folder Toolbar, a dialog box will appear giving you the option of changing the folder listed in the Current Folder Toolbar to the folder containing your script (see Figure 2.7). If a path to the folder containing the script has already been established, click on the Change Folder button.

7. If you need additional help on getting started, you can click on the *Help* icon ? in the Toolstrip in MATLAB's desktop (see Figure 2.2). If you are in the Editor Window, click on Home (upper left) to get back to MATLAB's desktop. In the window that opens (see Figure 2.8) you can type-in an item of interest in the search box, or you can click on the MATLAB option that brings up the window shown in Figure 2.9.

8. Whenever you write a script, it is good practice to add comment lines at the beginning of the script describing what the program is about. This is accomplished by placing a % sign in front of a statement in the script. Example:

```
% This program plots velocity vs. time.
```

16                                                                                    MATLAB® Essentials

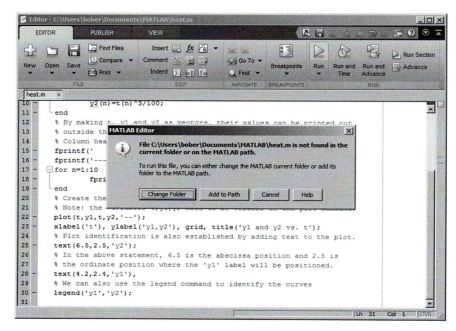

**FIGURE 2.7**
Dialog box for changing folder or path.

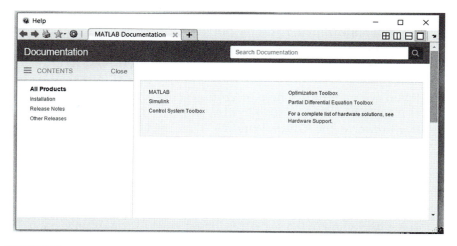

**FIGURE 2.8**
Help window.

# MATLAB® Fundamentals

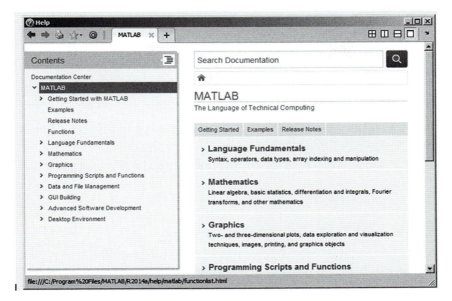

**FIGURE 2.9**
Topics in the MATLAB Help window.

## REVIEW 2.1

1. What are the two alternative ways to start the MATLAB program?
2. What are the windows in the MATLAB's default desktop?
3. It is best to write a MATLAB script (program) in the Editor Window. From MATLAB's default desktop, how does one open the Editor Window?
4. After you have completed writing a script in the appropriate window, what is the next step?
5. Name two ways to execute a script.
6. What happens if you try to run a script and the folder containing the script is not listed in the Current Folder Toolbar?
7. In MATLAB, what is the file name extension for saved scripts?
8. How does one establish a comment line in a script?

## 2.4 Variable Names and Types

- Variable names must start with a letter.
- Can contain letters, digits, and the underscore character (no spaces).
- Can be of any length, but must be unique within the first 19 characters.

**NOTE:** *Do not use a variable name that is same as a file name, a MATLAB function name or a self-written function name.*

A variable can be a scalar (A = 3.5), a vector (A = [2 4 6 8]), or a matrix

$$\left(A = \begin{bmatrix} 1 & 3 \\ 6 & 5 \end{bmatrix}\right)$$

A scalar has just one value in the computer memory, whereas a vector of $n$ elements will have $n$ values in the computer memory, and a matrix of $n$ rows and $m$ columns will have $n \times m$ values in the computer memory.

To make it easier to follow ones program, it is best to use variable names that are similar to the variables used in a problem statement.

MATLAB command names and variable names are case sensitive.

Numerical variables can be either an integer (no decimal point) or a floating point number (one with a decimal point). Integers can be stored in either 8 bits (numbers less than 127 or greater than −127), 32 bits (numbers less than 32,767 or greater than −32,767) or 64 bits of memory. In MATLAB, the default for floating point numbers is double precision that requires 64 bits of memory. You can specify variables to be single precision that only requires 32 bits of memory. Calculations carried out in single precision are faster than carrying out calculations in double precision. For all problems in this textbook, time of execution is not a problem. Numbers larger than approximately $3.4 \times 10^{38}$ or less than $-3.4 \times 10^{38}$ requires double precision.

When defining a variable, either in the Command Window or in a script, you should place a semicolon after the variable definition when you do not want the command echoed to the screen. In the absence of a semicolon, the defined variable appears on the screen. For example, if you entered the following command in the Command Window:

```
>> A = [3 4 7 6]
```

In the Command Window, you would see

```
A =
     3   4   7   6
>>
```

Alternatively, if you add the semicolon after the command statement, then your command is entered but there is nothing printed to the screen, and the prompt immediately appears for you to enter your next command:

```
>> A = [3 4 7 6];
>>
```

## 2.5 Assignment Operator

The assignment operator is of the form

$$\text{Variable name} = \text{an expression}$$

In most cases in this book the expression will be an *Arithmetic Statement* involving constants, Arithmetic Operators, variables, MATLAB functions and self-written functions. The one exception would be when we are dealing with characters and strings. Many MATLAB built-in functions that may be used in an arithmetic statement are discussed later. The way it works is that the Arithmetic Logic Unit in the computer will determine a value of the expression on the right-hand side of the = sign and replace the value of the variable on the left-hand side of the = sign with the value determined by the expression on the right-hand side of the = sign. For example,

Suppose you had the following assignments:

```
x=10; y=20;
x=x+y;
```

What the computer does is to determines the value of x+y, which equals 30, and replaces the original value of x, which is 10, with the new value of 30. Although, algebraically, the expression x=x+y does not make sense, since that would make y=0, it does make sense in programming language. Although, algebraically you can write x+y=20, you cannot do this in the programming language. There needs to be a single variable on the left-hand side of the = sign.

**NOTE:** *In an Arithmetic Statement, all variables on the right-hand side of the equal sign must be previously defined (given a numerical value) in the program.*

You may have noticed that in the variable assignments above that we placed a semicolon at the end of the assignment. This avoided the variables from being echoed to the screen.

### 2.5.1 Arithmetic Operators

The Arithmetic Operators used for addition, subtraction, multiplication, division, and exponentiation are listed below:

- \+ Addition
- \- Subtraction
- \* Multiplication
- / Division
- ^ Exponentiation

For Arithmetic Statements that contain several Arithmetic Operators and parentheses, there is a specific order that is used in evaluating the arithmetic expression. First, the Arithmetic Logic Unit, going from left to right, searches for parentheses, if it finds them, it will carry out the operations inside all the parentheses in the following order: exponentiation, multiplication and division, and addition and subtraction. It then returns to the beginning of the arithmetic statement and carries out the operations in the same order listed above.

Knowing this order may help you in deciding where parentheses are required when you write arithmetic statements. Suppose you had an expression $y = c/2m$, you might be tempted to write the expression in the MATLAB Command Window (after defining $c$ and $m$) as

```
clc;
c = 36.0; m = 3.0;
y = c/2*m
```

This would give the wrong answer for $y$. MATLAB would divide $c$ by 2 and multiply the result by $m$. The correct ways to write the expression are

```
y = c/(2*m)  or  c/2/m
```

In the first expression, MATLAB will first carry out the expression within the parentheses, so that the 2*$m$ becomes one number, and then $c$ is divided by this one number. In the second expression, there are no parentheses, so MATLAB, proceeding from left to right, will calculate $c/2$, then divide the result by $m$. Try typing these expressions in the Command Window and observe the two different answers you get for $y$.

To display a variable value, just type the variable name without the semicolon, and the variable will appear on the screen.

Try typing these commands into the Command Window and verify the results:

# MATLAB® Fundamentals

```
clc;
x = 5; y = 10; z = x + y <enter>
w = x - y <enter>
z = y/x <enter>
z = x*y <enter>
u = x^2 <enter>
```

---

## 2.6 Some MATLAB Features, Commands, Special Items, and Built-in Functions

### 2.6.1 Trigonometric and Other Useful Functions

Whenever you write a script, you should add comment lines to the script that describes what the script is about. You do this by entering the percent sign (%) at the beginning of the line. Example:

% This script determines the velocity of a free falling body ...

You may also add a comment after a particular command.

Whenever you write a script, it is a good practice to clear out variables that are left in the workspace from previous programs, since there could be a conflict between the variables used in the present program with those used in a previous program. You may also wish to clear contents in the Command Window, so that only results from the present program appear in the Command Window. You may accomplish this by placing the following commands at the beginning of your script (after your comment lines describing what your script is about).

    `clear;`   removes all variables and data from the work space.
    `clc;`     clears the Command Window.

If you wish to clear a graphics window, use

    `clf;`     clears the Graphics Window.

If you find that your program is running in an infinite loop, you can halt the program by hitting the ctrl and C keys simultaneously, that is,

    `ctrl-C`   aborts a program that may be running in an infinite loop.

Commands are case sensitive. Use lowercase letters for commands.

The `quit` command or `exit` command terminates MATLAB.

The `save` command saves variables or data in the work space of the Current Folder. The data file name will have the *.mat* extension.

The basic data structure in MATLAB is a matrix.

A matrix is surrounded by brackets and may have an arbitrary number of rows and columns; for example, the matrix

$$A = \begin{bmatrix} 1 & 3 \\ 6 & 5 \end{bmatrix}$$

may be entered into MATLAB as

```
>> A = [ 1 3 <enter>
         6 5 ]; <enter>
```

or

```
>> A = [ 1 3 ; 6 5 ]; <enter>
```

where the semicolon within the brackets indicates the start of a new row within the matrix. In the above expression for matrix $A$, row 1 are the elements 1 and 3, row 2 are the elements 6 and 5, column 1 are the elements 1 and 6, and column 2 are the elements 3 and 5.

A matrix of 1 row and 1 column is a scalar. Example:

```
>> A = [ 3.5 ];
```

Alternatively, MATLAB also accepts `A=3.5` (without brackets) as a scalar.

A matrix consisting of 1 row and several columns, or 1 column and several rows is considered a vector. Example:

```
>> A = [ 2 3 6 5 ]   (row vector)
>> B = [ 2
         3
         6
         5 ]   (column vector)
```

We can convert a column vector to a row vector by using the transpose symbol. Suppose we enter **B** as a column vector in the Command Window then write **B'**. We would see the following in the Command Window:

```
>> B=[2
      3
      6
      5];
>> B'
ans =
      2    3    6    5
>>
```

A matrix can be defined by including a second matrix as one of the elements. Example:

```
>> B = [ 1.5 3.1 ];
>> C = [ 4.0 B ];   (thus C = [ 4.0 1.5 3.1] )
```

You can select a specific element of the vector c as follows:

If C = [ 4.0 1.5 3.1], then
```
>> b = C(2)
```

gives $b = 1.5$.

If $A = \begin{bmatrix} 1 & 3 \\ 6 & 5 \end{bmatrix}$, then
```
>> b = A(2,2)
```
gives $b = 5$.

You can also define a vector by specifying each element in the vector, for example:

```
A(1)=1, A(2)=3, A(3)= 5, A(4)=7, then vector A=[1 3 5 7].
```

*The element number must be an integer.*

This concept is very important and is used in many examples that follow.

### 2.6.1.1 Special Values

One special value in MATLAB is ans, it is the last computed unassigned result to an expression typed in the Command Window. For example, if we typed in the following assignments in the Command Window, we can see MATLAB's response.

```
>> x=5; y=10;
>> x+y
ans =
     15
>>
```

MATLAB has a built-in value for the variable $\pi$. Its symbol is pi, and it should be used in place of 3.14 whenever $\pi$ appears in an arithmetic statement.

Typing pi in the Command Window gives

```
>> pi
ans =
    3.1416
>>
```

The display default is four places, but it is carried to many more places in memory.

If you had an expression in which you accidentally divided by zero, MATLAB would respond with the infinity value, $\infty$, with the symbol, inf. Example:

```
>> x=10; y=0;
>> z=x/y
z =
    Inf
>>
```

### 2.6.1.2 Trigonometric Functions

There are many engineering examples that involve the trigonometric functions. Similar to other computer programs, MATLAB has functions that evaluate the trigonometric functions. The most frequently used are shown below.

| MATLAB's Function | Trigonometric Function Name |
|---|---|
| sin( ) | sine |
| cos( ) | cosine |
| tan( ) | tangent |

The arguments of these trigonometric functions are in radians. However, the arguments can be made in degrees if a d is placed after the function name, such as sind(x). In all of the trigonometric functions, you may use simple arithmetic in the arguments of the function.

Try typing these statements in the Command Window and use your calculator to verify the results.

```
clc;
x = 50/180*pi; y = sin(x)   <enter>
z = cos(pi/2)   <enter>
```

The answer should be 0, but with round off error it gives 6.1232e-17.

```
w = tan(pi/4)   <enter>
x = 45/180*pi; y2 = sin(x)   <enter>
z2 = cos(x)   <enter>
x1 = sind(50)   <enter>
y1 = cosd(90)   <enter>
w1 = tand(45)   <enter>
```

### 2.6.1.3 Inverse Trigonometric Functions

| MATLAB's Function | Trigonometric Function Name |
|---|---|
| asin( ) | Inverse sine |
| acos( ) | Inverse cosine |
| atan( ) | Inverse tangent |

Since the values of the sine and cosine functions vary from −1 to +1. The input arguments to the asin( ) and acos( ) functions should be from −1 to +1. The results will be in radians. The value of the tan function can be anywhere from ($-\infty$ to $+\infty$), so the input argument to the atan( ) function can be any number, but the result will be in radians.

# MATLAB® Fundamentals

Try typing these statements into the Command Window and use your calculator to verify the results:

```
clc;
x = asin(0.5); xd = x*180/pi  <enter>
y = acosd(-1.0)  <enter>
z = atand(1.732)  <enter>
z = atan(1.0); zd = z*180/pi  <enter>
```

### 2.6.1.4 Exponential, Square Root, and Error Functions

| MATLAB's Function | Mathematics Function Name |
|---|---|
| exp( ) | Exponential ($e^{()}$, $e \approx 2.7183$) |
| log( ) | Natural logarithm |
| log10( ) | Common (base 10) logarithm |
| sqrt( ) | Square root |
| erf( ) | Error function |

Try typing these statements into the Command Window and use your calculator to verify the results:

```
clc;
x = 2.5; y = exp(x)  <enter>
z = log(y)  <enter>
w = sqrt(x)  <enter>
u = log10(100)  <enter>
```

Suppose we had a problem involving the following arithmetic statement that we needed to evaluate:

$$y = \cos\left(\sqrt{\frac{k}{m} - \left(\frac{c}{2m}\right)^2}\, t\right)$$

To make it easier to write the MATLAB statement corresponding to the above arithmetic statement, we could break up the argument of the cos function as follows (type the following in the Command Window):

```
k = 200; c = 5; m = 25; t = 5;
arg = sqrt(k/m - (c/(2*m))^2);
y = cos(arg*t )
```

### 2.6.1.5 Complex Numbers

Complex numbers may be written in two forms: Cartesian, for example:

```
z = x + yj;
```

The x part is considered the real part of the complex number and the y part is considered as the imaginary part of the complex number.

The complex number can also be expressed in polar form, for example:

```
z = r * exp(j*theta).
```

MATLAB allows the use of i and j for $\sqrt{-1}$. Programmers who have experience with FORTRAN, the programming language that was commonly used in engineering many years ago, frequently used i and j as **integer** loop variables.

In this book we do not deal with complex numbers very often, but when we do, we will use j for $\sqrt{-1}$. Also, there are many examples in this book where i is used as an integer loop variable.

### 2.6.2 Other Special Values

| MATLAB's Function | Math Function Name |
|---|---|
| abs( ) | Absolute value (magnitude) |
| conj( ) | Complex conjugate |
| imag( ) | Imaginary part of a complex number |
| real( ) | The real part of a complex number |

Try typing these statements into the Command Window and use your calculator to verify the results:

```
clc;
z1 = 1 + j; z2 = 2*exp(j*pi/6)=2*(cos(pi/6)+j sin(pi/6));
y = abs(z1) <enter>
w = real(z2) <enter>
v = imag(z2) <enter>
```

### 2.6.2.1 Other Useful MATLAB Functions

| | |
|---|---|
| size(X) | Gives the size (number of rows and the number of columns of matrix X). |
| x' | Transposes a matrix or vector, rows become columns and columns become rows. |
| length(X) | For vectors, length(X) gives the number of elements in X. |

*(Continued)*

# MATLAB® Fundamentals

| | |
|---|---|
| `linspace(X1,X2,N)` | Generates N points between X1 and X2. |
| `sum(X)` | For vectors, `sum(X)` gives the sum of the elements in X. For matrices, `sum(X)` gives a row vector containing the sum of the elements in each column of the matrix. |
| `max(X)` | For vectors, `max(X)` gives the maximum element in X. For matrices, `max(X)` gives a row vector containing the maximum in each column of the matrix. If X is a column vector, it gives the maximum absolute value of X. |
| `min(X)` | Same as `max(X)` except it gives the minimum element in X. |
| `mean(X)` | The mean of a vector, also known as the average, equals the sum of the vector elements divided by the number of elements in the vector. For vectors, `mean(X)` gives the mean value of the vector X. For matrices, `mean(X)` gives a row vector containing the mean value in each column of the matrix X. |
| `sort(X)` | For vectors, `sort(X)` sorts the elements of X in ascending order. For matrices, `sort(X)` sorts each column in the matrix in ascending order. |
| `factorial(n)` | $n! = 1 \times 2 \times 3 \times ... \times n$ |
| `mod(x,y)` | Modulo operator gives the remainder resulting from the division of x by y. For example, `mod(13,5) = 3`, that is, $13 \div 5$ gives 2 plus remainder of 3 (2 is discarded). Another example: `mod(n,2)` gives zero if n is an even integer and one if n is an odd integer. |

Try typing these statements into the Command Window:

**NOTE:** The repeat of A is not necessary, but it will make it easy to see the results of n, y, z, w, and u.

```
clc;
A = [ 2 15 6 18 ]; n = length(A) <enter>
A <enter> y = max(A) <enter>
A <enter> z = sum(A) <enter>
A <enter> w = mean(A) <enter>
A <enter> u = sort(A) <enter>
A = [ 2 15 6 18; 15 10 8 4; 10 6 2 3 ]; <enter>
A' <enter>
A <enter> x = max(A) <enter>
A <enter> y = sum(A) <enter>
A <enter> w = mean(A) <enter>
A <enter> u = sort(A) <enter>
A <enter> z = size(A) <enter>
w = mod(21,2) <enter>
u = mod(20,2) <enter>
```

A list of the complete set of elementary math functions can be obtained by typing **help elfun** in the Command Window.

### 2.6.2.2 Colon Operator (:)

The colon operator may be used to

1. Create a new matrix from an existing matrix; examples:

$$\text{if } A = \begin{bmatrix} 5 & 7 & 10 \\ 2 & 5 & 2 \\ 1 & 3 & 1 \end{bmatrix}$$

$$\text{then } x = A(:,1) \text{ gives } x = \begin{bmatrix} 5 \\ 2 \\ 1 \end{bmatrix}$$

The colon in the expression A(:,1) implies all the rows in matrix $A$, and 1 implies column 1.

$$x = A(:,2:3) \text{ gives } x = \begin{bmatrix} 7 & 10 \\ 5 & 2 \\ 3 & 1 \end{bmatrix}$$

The first colon in the expression A(:,2:3) implies all the rows in $A$, and the 2:3 implies columns 2 and 3.

We can also write

$$y = A(1,:) \text{ that gives } y = [5 \ 7 \ 10]$$

The 1 implies the first row and the colon implies all the columns.

2. Colon operator can also be used to generate a series of numbers (as in a `for` loop, which is discussed later) or to create a vector. The format is

$n$ = starting value: step size: final value.

If the step size is omitted, the default step size is one. Example:
$n$ = 1:8 gives $n = \begin{bmatrix} 1 & 2 & 3 & 4 & 5 & 6 & 7 & 8 \end{bmatrix}$.
To increment in steps of 2 use
$n$ = 1:2:7 gives $n = \begin{bmatrix} 1 & 3 & 5 & 7 \end{bmatrix}$

### Exercise

**E2.1.** Type the following matrix in the Command Window. Assume that the first, second, and third columns represent the vector variables of altitude, $z$, temperature, $T$, and density, *rho* respectively: (a) use the colon operator to define the vector variables, (b) determine the mean values of altitude, temperature, and density, (c) determine the length of vector $z$, and (d) determine the size of matrix $A$. Print the results to the Command Window.

$$A = \begin{bmatrix} 0 & 288.15 & 1.2252 \\ 1000 & 281.65 & 1.1118 \\ 2000 & 275.15 & 1.0065 \\ 3000 & 268.65 & 0.9091 \\ 4000 & 262.15 & 0.8191 \\ 5000 & 255.65 & 0.7360 \end{bmatrix}$$

### 2.6.2.3 Preallocation of a Matrix

Sometimes it is necessary to preallocate a matrix of a given size. This can be done by defining a matrix of all zeros or ones; Examples:

$$A = \text{zeros}(3) = \begin{bmatrix} 0 & 0 & 0 \\ 0 & 0 & 0 \\ 0 & 0 & 0 \end{bmatrix} \text{ (3 rows, 3 columns)}$$

$$B = \text{zeros}(3:2) = \begin{bmatrix} 0 & 0 \\ 0 & 0 \\ 0 & 0 \end{bmatrix} \text{ (3 rows, 2 columns)}$$

---

**REVIEW 2.2**

1. List at least two conditions in selecting a name for a variable.
2. Finish the following statement. An arithmetic statement may involve ...
3. What can be said about the variables that appear on the right side of an arithmetic statement?
4. List the Arithmetic Operators in MATLAB.
5. What is the order in which an arithmetic statement will be carried out?

*(Continued)*

### REVIEW 2.2 (Continued)

6. What is MATLAB's command for
   a. $\pi$.
   b. $e$.
   c. ln.
   d. Sine function in radians.
   e. Sine function in degrees.
   f. $\sin^{-1}$ function.
   g. The number of elements in a vector.
   h. The size of a matrix (the number of rows and columns).
   i. The sum of the elements in a vector.
   j. The maximum element in a vector.
   k. Preallocating the size of a 3 × 3 matrix.
7. What is the purpose of placing a semicolon at the end of a command statement or a variable assignment?
8. What is the command that will clear the Command Window?
9. What is the basic data structure in MATLAB?
10. Name two functions of the colon operator.

## 2.7 MATLAB Output

To display a vector X, just type X without the semicolon, and vector X will be printed to the screen. For example, first define X,

```
>> X = [0 1 2 3 4 5];
>>
```

Now enter X without the semicolon.

```
>> X
```

The following will be displayed on the screen:

```
X =
       0     1     2     3     4     5
>>
```

## 2.7.1 The `disp` Command

The `disp` command prints only the items that are enclosed within the parentheses, which can be a variable or alphanumeric information. Alphanumeric information must be enclosed by singe quotation marks. Example (assuming that vector X above has already been entered in the Command Window) type in

```
disp(X); disp(' m/s');
```

The following will be displayed on the screen:

```
     0    1    2    3    4    5
m/s
>>
```

## 2.7.2 The `fprintf` Command

The `fprintf` command prints formatted text to the screen or to a file.
  Example:

```
>> V = 2.2; clc;
fprintf('The velocity is %f m/s \n', V);
```

The following will appear on the screen:

```
The velocity is 2.200000 m/s
```

The \n in the above command tells MATLAB to move the cursor to the next line.
  MATLAB also has a tab command. It is \t; this command tells MATLAB to move the cursor several spaces along the same line.
  The %f refers to a formatted floating point number that is assigned to variable V, and the default is 6 decimal places. The command `fprintf` uses format strings based on the C programming language. You can specify the number of spaces allotted for the printed variable as well as the number of decimal places by using %8.2f. This will allow eight spaces for the variable to two decimal places. You can also just specify the number of decimal places for the variable and let MATLAB decide the number of spaces allotted for the printed variable. For example, to specify three decimal places use %.3f. The variable will be printed out to three decimal places, but MATLAB will decide the number of spaces for the variable. However, to create neat looking tables, it is best to specify the number of spaces in the format statement that allows for several spaces between variables in adjacent columns, such as %10.3f.

Other formats:

| | |
|---|---|
| %i or %d | Used for integers |
| %f | Used for a floating point number (one with a decimal point) |
| %e | Scientific notation (e.g., 6.02e23), default is 6 decimal places |
| %g | Automatically uses the briefest of %f or %e format |
| %s | Used for a string of characters |
| %c | Used for a single character |

Unlike C, the format string in MATLAB's `fprintf` must be enclosed by *single quotation marks (and* not *double quotes).*

### 2.7.3 Printing to a File

It is often useful to print the results of a MATLAB program to a file, possibly for inclusion in a report. In addition, program output that is printed to a file can be subsequently edited within the file, such as aligning or editing column headings in a table. Before you can print to a file, you need to open a file for printing with the command, `fopen`. The syntax for `fopen` is

```
fo = fopen('filename','w')
```

Thus, `fo` is a pointer to the file named `filename`, and the w indicates that there will be writing to the file. The `fo` can be replaced by a name selected by the programmer. To print to `filename` use

```
fprintf(fo,'format',var1,var2,..);
```

where the `format` string contains the text format for `var1`, `var2`, and so on.

Try typing the following example script in the Editor Window, save the script in the folder that you have chosen for your MATLAB scripts (this becomes the *current folder*), then run the script (see Section 2.3). The output should go to the file named *output.txt*, which should be located in the same folder as the script that produced it. To see the results, open the output file, as described below. If you wish, you can edit the results and also print the results by clicking on the *print* command in the Toolstrip.

**Example 2.1**

```
% Example_2_1.m
% This program is an example for printing to a file.
clear; clc;
V=12; % velocity
F=50.2; % force
fo=fopen('output.txt','w');
fprintf(fo,'V=%4i m/s, F = %5.2f N \n',V, F);
fclose(fo);
```
---

## MATLAB® Fundamentals

**Program Results:**
```
V= 12 m/s, F = 50.20 N
```
--------------------------------------------------------------

The extension on the output file should be *.txt* (otherwise when you try to open the file, MATLAB will start the import wizard). The resulting output file will be saved and listed in the Current Folder. You can open the file by double clicking on the *output.txt* file listed in the Current Folder column on the left (see Figure 2.10). Alternatively, you can open the file by clicking on the *Open* icon in the Toolstrip that brings up a screen listing all the *.m* files in the *Current folder*. In the box labeled *File name*, type in *\*.txt*. This will bring up a screen listing of all the files with the extension *.txt* in the *Current folder* as shown in Figure 2.11. To open the file of interest, double click on the name of the output file (in this example, the file name is *output.txt*).

In earlier versions of MATLAB, you would not be able to open the output file without having included the `fclose(fo)` statement in the program. But it is still a good practice to include the `fclose` statement after all the output statements in the program, or at the end of the program itself.

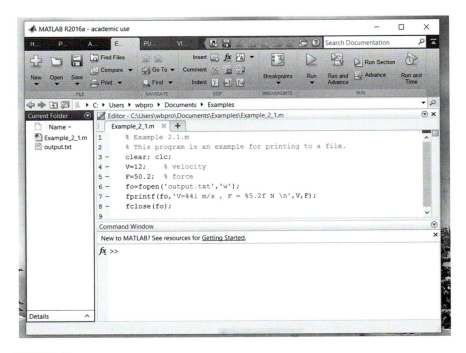

**FIGURE 2.10**
Searching for the output.txt file in Current Folder Column.

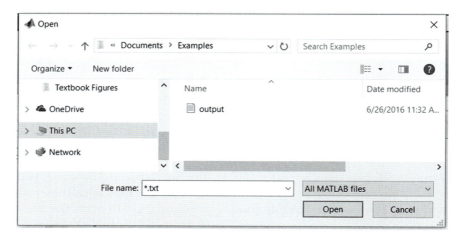

**FIGURE 2.11**
Opening all files with .txt extension.

## 2.8 Simple Plot Commands

MATLAB provides many different types of plots. Clicking on the PLOTS tab in MATLAB's desktop graphically lists the various types of plots that are available (see Figure 2.1). The commands for creating linear plots, semilog plots, and log-log plots are as follows:

| | |
|---|---|
| `Plot(x,y)` | Linear plot of $y$ versus $x$ |
| `Semilogx(x,y)` | Semilog plot (log scale for $x$-axis, linear scale for $y$-axis) |
| `Semilogy(x,y)` | Semilog plot (linear scale for $x$-axis, log scale for $y$-axis) |
| `Loglog(x,y)` | Log-log plot (log scale for both $x$- and $y$-axes) |

*Unless you wish to plot a single point, the arguments in the* `plot` *command must be vectors.* In addition, the vectors need to be of the same length. If the arguments in the plot command are scalars, the plot commands will produce just a single point.

### 2.8.1 Linear Plot

Suppose we have a relationship of $V = f(t)$ and we have created the following vectors (V(1) occurs at $t$(1), V(2) occurs at $t$(2), etc.).

$$t = \begin{bmatrix} 0.0 & 0.5 & 1.0 & 1.5 & 2.0 & 2.5 & 3.0 & 3.5 & 4.0 & 4.5 & 5.0 \end{bmatrix}$$

$$V = \begin{bmatrix} -20.2 & -21.0 & -19.4 & -14.7 & -6.2 & 6.9 & 25.4 & 50.0 & 81.4 & 120.4 & 167.8 \end{bmatrix}$$

# MATLAB® Fundamentals

In the following script we plot V versus *t* using the `plot(t,V)` command. We will assume that *t* is in seconds and V is in meters/second.

We can label the *t*-axis, *v*-axis, and add a title and a grid with the following commands:

```
xlabel('t(s)'),
        ylabel('V(m/s)'),
title('V vs. t'),
grid;
```

**Example 2.2**

```
% Example_2_2.m
% The vectors t and V are entered into the program.
% Then a plot of V vs. t is created.
% To plot V vs. t both variables need to be vectors
% of the same length.
clear; clc;
t = [0.0  0.5  1.0  1.5  2.0  2.5  3.0  3.5  4.0  4.5  5.0];
V = [-20.2 -21.0 -19.4 -14.7 -6.2 6.9 25.4 50.0 81.4 120.4...
     167.8];
% Create the plot of V vs. t.
plot(t,V), xlabel('t(s)'), ylabel('V(m/s)'), title('V vs. t'), grid;
```

**Program Results:**

See Figure 2.12.

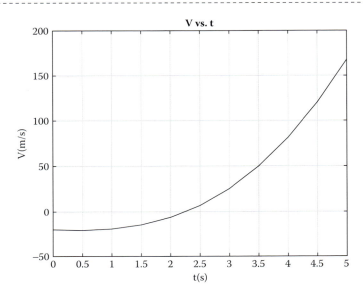

**FIGURE 2.12**
Plot of V versus t.

> **REVIEW 2.3**
>
> 1. Name two commands that will result in printing to the screen.
> 2. What is the command that will move the cursor to the next line?
> 3. What is the format that will print a floating point variable to 10 spaces and to three decimal points?
> 4. What is the format that will print a floating point variable in scientific notation to 12 spaces and to four decimal points?
> 5. What are the commands necessary to print to a file?
> 6. What is the command to create a plot of $y$ versus $x$ and what type of variable must $x$ and $y$ be?
> 7. What are the commands that will label the $x$- and $y$-axis and provide a title to a plot?

If a program involves creating more than one plot, you need to include the statement `figure` after each plot command (except the last), otherwise only the last plot will appear.

You may have noticed that the script for Example 2.2, contained comment lines that described what the script is about. This is a good practice and should be implemented every time you write a script.

## 2.9 Loops

### 2.9.1 The for Loop

The `for` loop command provides the means to *repeat* a series of statements with just a few lines of code. In MATLAB, in many cases, one can avoid the use of the `for` loop and achieve the same result. However, the method used in MATLAB to achieve this may not be available in many other computer platforms. Since we are emphasizing the computer programming building blocks that are applicable in most, if not all, programming languages, we will exclusively use the `for` loop method in the first few chapters of this book.

Syntax:

```
for loop variable = starting value: step size: final value
```

The step size may be omitted, and then MATLAB will take the step size to be 1. Although the loop variable need not be an integer, in most cases in this book, it will be an integer. That is because, we frequently use the loop

## MATLAB® Fundamentals

variable to select or create an element of a vector or a matrix. Elements of a vector are identified by an index which must be an integer. In most other platforms, if you wish a variable to be an integer, you need to designate that variable as an integer. This is not the case in MATLALB. MATLAB looks at the context in which the variable is used and knows when to consider the variable as an integer.

As an example, we will take the index variable as m, the starting value as 1, omit the step size and take the final value as 20, then our for loop will be

```
for m = 1:20
    statement;
         .
         .
         .
    statement;
end
```

MATLAB sets the index m to 1, carries out the statements between the for and end statements, then returns to the top of the loop, changes m to 2 and repeats the process. After the process has been carried out 20 times the program exits the loop without further executing any of the statements within the loop.

**NOTE:** *There is no semicolon after the* for *and* end *statements.*

*All statements that are not to be repeated should not be within the* for *loop. For example, table headings that are not to be repeated should be outside the* for *loop. Also notice that statements within the* for *loop are indented. MATLAB does this to make it easier to read and debug a script containing* for *loops. You can have MATLAB to do final indenting by highlighting your entire script and then entering Ctl-I.*

### Example 2.3

```
% Example_2_3.m
% This program is an example of the use of a for loop in which
% the indices of the for loop select an element of a vector.
% The indices must be an integer. In the for loop expression,
% MATLAB will take i as an integer if there is no decimal point
% in the assignment of variable i. But you can also specify i
% by the assignment int8(i) or int32(i) before it is used.
% The assignment length(t) specifies the number of elements
% in the vector t.
% The program creates a table of y1 and y2 vs t.
% 0 <= t <= 10 in steps of 0.5.
clear; clc;
% Table headings:
fprintf('    t          y1           y2   \n');
fprintf('-------------------------------\n');
t=0:0.5:10;
```

```
for i=1:length(t)
    y1=t(i)^2/10;
    y2=t(i)^3/100;
    fprintf('%5i       %10.3f       %10.3f \n',i,y1,y2);
end
```

**Program Results:**

```
    t              y1              y2
---------------------------------------
   0.0           0.000           0.000
   0.5           0.025           0.001
   1.0           0.100           0.010
   1.5           0.225           0.034
   2.0           0.400           0.080
    .              .               .
    .              .               .
   8.0           6.400           5.120
   8.5           7.225           6.141
   9.0           8.100           7.290
   9.5           9.025           8.574
  10.0          10.000          10.000
>>
```

In the above example, we selected an element of vector t by the loop variable i. But we did not make y1 and y2 as vectors. Thus, we would not be able to plot y1 and y2 versus t.

When there is a large output to the Command Window, you might wish to separate the Command Window from the Editor Window. You can do this by clicking on the down arrow within the little circle in the black section of the Command Window and selecting the undock option in the dropdown window (see Figure 2.13).

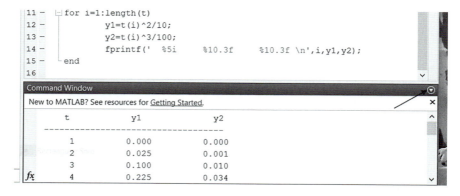

**FIGURE 2.13**
How to undock Command Window from Editor Window?

## MATLAB® Fundamentals

In the next example the loop variable is not an integer, and thus we would not be able to use the loop variable to select an element of a vector.

**Example 2.4**

```
% Example_2_4.m
% In this example the loop variable is x which is not an integer.
% Thus, the loop variable x could not be used to select an element
% of a vector. The range of x is from -0.9 to +0.9 in steps of 0.1.
clear; clc;
% print the table headings outside of the 'for' loop:
fprintf(' x           y1             y2  \n');
fprintf('-------------------------------------\n');
for x = -0.9:0.1:0.9
    y1=x/(1-x);
    y2=y1^2;
    fprintf('%5.2f      %10.3f      %10.3f \n',x,y1,y2);
end
% fprintf('\n %5.2f \n',x)
```

**Program Results:**

```
  x            y1           y2
-------------------------------
-0.90        -0.474        0.224
-0.80        -0.444        0.198
-0.70        -0.412        0.170
-0.60        -0.375        0.141
-0.50        -0.333        0.111
  .             .            .
  .             .            .
  .             .            .
 0.50         1.000        1.000
 0.60         1.500        2.250
 0.70         2.333        5.444
 0.80         4.000       16.000
 0.90         9.000       81.000
>>
```

You might think that the statement for x = -0.9:0.1:0.9 would create a vector x. However, that is not the case. The process starts by setting x = -0.9. As the program progresses back to the start of the for loop, old values of x are overwritten by the new value of x. Try adding the statement

```
fprintf(' %5.2f \n',x);
```

at the end of the for loop (by removing the % sign before the fprintf statement in the above program) and rerunning the program. See that you only get the last value of x, which is 0.9. Now type x =-0.9:0.1:0.9 in the Command Window without the semicolon. See that x is now a vector.

**Example 2.5**

In this example we will calculate the position and velocity of a free falling body in a gravitational field (neglecting drag) as a function of time, t. See Figure 2.14. The governing equations are based on Newton's second law and can be found in any university physics textbook.

The governing equations are

$$V = V_o - gt \tag{2.1}$$

$$y = V_o t - \frac{gt^2}{2} \tag{2.2}$$

where:
  V is the velocity
  y is the position pointing upward
  g is the acceleration of gravity
  t is the time

The following MATLAB program calculates and plots V, and y versus t, for $0 \le t \le 2$ s in steps of 0.1 s. We have taken $V_o = 10$ m/s, and $g = 9.81$ m/s². We will print a table consisting of t, V, and y at every other time step. In addition, we will plot V versus t and y versus t. Finally, we will determine the approximate maximum height reached by the free falling body.

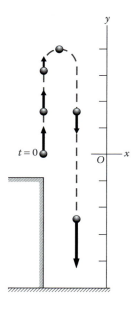

**FIGURE 2.14**
Ball in a gravitational field.

# MATLAB® Fundamentals

```
% Example_2_5.m
% This program calculates the velocity and position of a free
% falling body vs. time.
% The velocity, V = Vo-gt
% The position, y = Vo*t-0.5*g*t^2
% The initial velocity, Vo=10 m/s, g=9.81 m/s^2
% The output goes to a file named output.txt.
% Plots of y vs. t and V vs. t are made.
% The approximate maximum height reached by the body is determined.
clear; clc;
Vo=10.0; g=9.81;
fo=fopen('output.txt','w');
t=0:0.1:2;
for i=1:length(t)
    V(i)=Vo-g*t(i);
    y(i)=Vo*t(i)-0.5*g*t(i)^2;
end
plot(t,V), xlabel('t(s)'), ylabel('V(m/s)'), title('V vs. t'), grid;
figure;
plot(t,y), xlabel('t(s)'), ylabel('y(m)'), title('y vs. t'), grid;
ymax=max(y);
fprintf(fo,'The approximate maximum height reached by the body =');
fprintf(fo,' %8.3f m \n',ymax);
% Table headings
fprintf(fo,'t(s)     V(m/s)    y(m)   \n');
fprintf(fo,'-------------------------\n');
for i=1:2:length(t)
    fprintf(fo,'%6.2f     %10.2f     %10.2f \n',t(i),V(i),y(i));
end
```
---

## Program Results:

```
The approximate maximum height reached by the body = 5.095 m
t(s)    V(m/s)    y(m)
-------------------------
0.00    10.00    0.00
0.20     8.04    1.80
0.40     6.08    3.22
0.60     4.11    4.23
0.80     2.15    4.86
1.00     0.19    5.09
1.20    -1.77    4.94
1.40    -3.73    4.39
1.60    -5.70    3.44
1.80    -7.66    2.11
2.00    -9.62    0.38
```
---
See Figures 2.15 and 2.16.

---

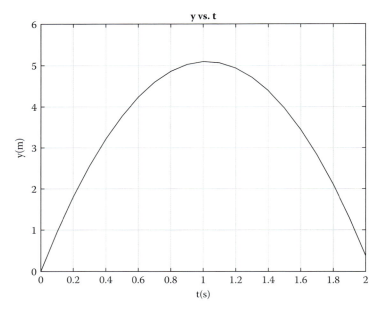

**FIGURE 2.15**
y versus t for ball in a gravitational field.

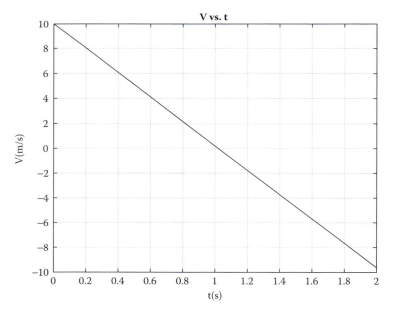

**FIGURE 2.16**
V versus t for ball in a gravitational field.

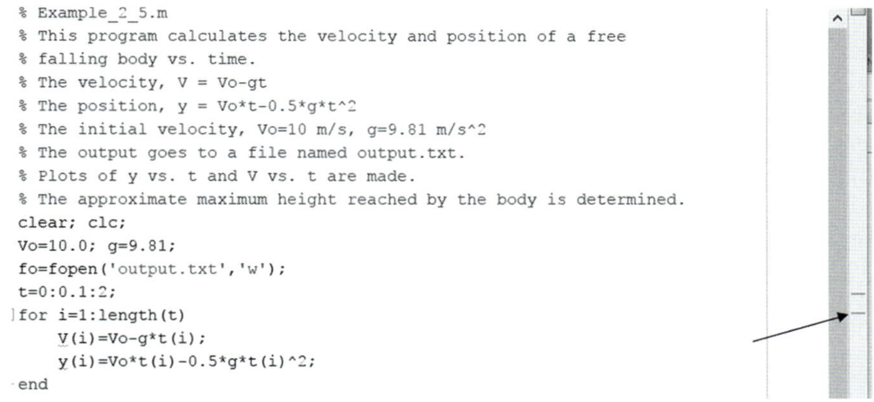

# MATLAB® Fundamentals

```
% Example_2_5.m
% This program calculates the velocity and position of a free
% falling body vs. time.
% The velocity, V = Vo-gt
% The position, y = Vo*t-0.5*g*t^2
% The initial velocity, Vo=10 m/s, g=9.81 m/s^2
% The output goes to a file named output.txt.
% Plots of y vs. t and V vs. t are made.
% The approximate maximum height reached by the body is determined.
clear; clc;
Vo=10.0; g=9.81;
fo=fopen('output.txt','w');
t=0:0.1:2;
for i=1:length(t)
    V(i)=Vo-g*t(i);
    y(i)=Vo*t(i)-0.5*g*t(i)^2;
end
```

**FIGURE 2.17**
Preallocation message when running Example 2.5.

In running Example 2.5, you may have noticed a small orange line just to the right of the vertical ladder (see Figure 2.17). If you use your mouse to point on the orange line you would get the following message: "The variable 'V' appears to change size with every loop iteration (within a script). Consider preallocating for speed." This would be very important when the number of repeats in the loop is very large; otherwise, it is not important. Although MATLAB recommends, but does not require, the preallocation of the size of the vector or matrix that is being generated, other programs such as C/C++ do require it. To preallocate the size of the vector that is being generated, use MATLAB's zeros function. In the above example, 21 v values and 21 y values will be generated. So add the following statements before the `for` loop:

`v=zeros(21,1)` and `y=zeros(21,1)`.

## 2.9.2 The While Loop

An alternative to the `for` loop is the `while` loop. If an index in the program is required, the use of the while loop statement (unlike the `for` loop statement) requires that the program generate its own index, as shown in the following example:

```
n = 0;
while n < 10
    n = n+1;
    y = n^2;
end
```

In the `while` loop, MATLAB will carry out the statements between the `while` and `end` statements as long as the condition in the `while` statement is satisfied. In the above example, when n = 10, none of the commands within the while loop will be executed and the program goes to the next command after the end statement. Note that the statement "n = n+1" above does not make sense algebraically, but does makes sense in the MATLAB language. *The "=" operator in MATLAB (as in many computer languages) is the* assignment operator that tells MATLAB to fetch the contents in the memory cell containing the variable n, put its value into the arithmetic unit of the CPU, increment the variable n by 1, and put the new value back into the memory cell designated for the variable n. Thus, the old value of n has been replaced by the new value for n.

In this example, we will use a simpler version of Example 2.5, but this time we will use the `while` loop instead of the `for` loop.

**Example 2.6**

```
% Example_2_6.m
% This program calculates the velocity and position of a free
% falling body vs. time.
% The program uses a while loop in place of the for loop.
% The velocity, V = Vo-gt
% The position y = Vo*t-0.5*g*t^2
% Vo=10 m/s, g=9.81 m/s^2
% The output goes to a file named output.txt.
clear; clc;
Vo=10.0; g=9.81;
fo=fopen('output.txt','w');
% Table headings
fprintf(fo,'t(s)     V(m/s)      y(m)    \n');
fprintf(fo,'-------------------------\n');
t=0; V=0; y=0;
while t<=2
    fprintf(fo,'%6.2f    %10.2f    %10.2f \n',t,V,y);
    t=t+0.2;
    V=Vo-g*t;
    y=Vo*t-0.5*g*t^2;
end
```
---

**Program Results:**

```
t(s)   V(m/s)    y(m)
-------------------------
0.00    10.00    0.00
0.20     8.04    1.80
0.40     6.08    3.22
0.60     4.11    4.23
0.80     2.15    4.86
1.00     0.19    5.09
1.20    -1.77    4.94
1.40    -3.73    4.39
1.60    -5.70    3.44
1.80    -7.66    2.11
2.00    -9.62    0.38
```
---

Compare results obtained from Examples 2.7 and 2.8.
Are they the same?

## REVIEW 2.4

1. What is the objective in using a `for` loop?
2. What is the syntax of a `for` loop?
3. Should table headings that are not to be repeated be inside a `for` loop?
4. If the index of a `for` loop is used to select an element of a vector or a matrix, what variable type should the `for` loop index be?
5. What other statement type can be used to create a loop?
6. What is the major difference between a `for` loop and a `while` loop?

## Exercises

**E2.2.** The motion of a piston in an internal combustion engine is shown in Figure 2.18a and b.

The piston's position, $s$, as seen from the crankshaft center is determined to be

$$s(t) = r\cos(2\pi\omega t) + \sqrt{b^2 - r^2 \sin^2(2\pi\omega t)} \qquad (2.3)$$

where:
$b$ is the length of the piston rod
$r$ is the radius of the crankshaft
$\omega$ is the rotational speed of the crankshaft in revolutions per second

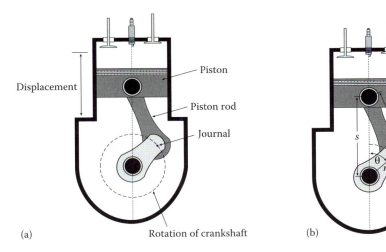

**FIGURE 2.18**
(a) Piston configuration and (b) piston position variables.

Develop a MATLAB program that determines s versus t for $0 \leq t \leq 0.01$ s. Use 20 subdivisions on the t domain. Take $r = 9$ cm, $\omega = 100$ revolutions per second, and $b = 14$ cm.

1. Create a table of s versus t and print the results to both the screen and to file.
2. Create a plot of s versus t.

**E2.3.** The position, $y$, of a mass in a mass-spring-dashpot system (see Figure 2.19) is given by

$$y = \exp\left(-\frac{c}{2m}t\right)\left\{A\sin\left(\sqrt{\frac{k}{m} - \left(\frac{c}{2m}\right)^2}\,t\right) + B\cos\left(\sqrt{\frac{k}{m} - \left(\frac{c}{2m}\right)^2}\,t\right)\right\} \quad (2.4)$$

Take
$m = 25.0$ kg
$c$ is the damping factor $= 5.0$ N-s/m
$k$ is the spring constant $= 200.0$ N/m
$A = 5.0$ m
$B = 0.25$ m

1. Determine $y(t)$ for $0 \leq t \leq 10$ seconds in steps of 0.1 seconds.
2. Create a table of $y$ versus $t$ every 1 second and print the results to the screen.
3. Create a plot of $y$ versus $t$.

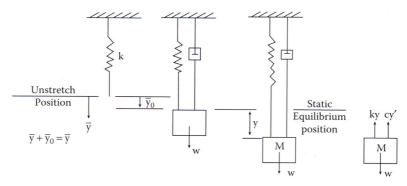

**FIGURE 2.19**
Mass-spring-dashpot system.

# MATLAB® Fundamentals

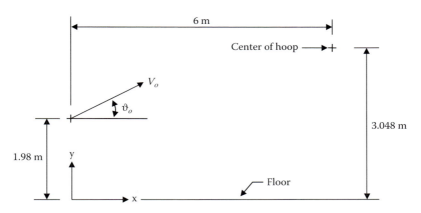

**FIGURE 2.20**
Basketball player shooting the basketball.

**E2.4.** A basketball player shoots the ball when he is 6 m from the center of the hoop as shown in Figure 2.20. The ball is released at a velocity, $V_o$, and makes angle $\vartheta_o = 40°$ with the horizontal. Using Newton's second law and the initial conditions and neglecting the drag on the basketball, we can determine the following equations for the $(x, y)$ position of the ball as a function of time, $t$.

$$x = V_o \cos(\vartheta_o) t \qquad (2.5)$$

$$y = y_o + V_o \sin(\vartheta_o) t - \frac{g}{2} t^2 \qquad (2.6)$$

Take the $(x, y)$ position of the center of the hoop to be $(x_f, y_f) = (6.0 \text{ m}, 3.048 \text{ m})$, $y_o = 1.98$ m, and $\vartheta_o = 40°$.

1. Determine the time, $t_f$, which it takes for the ball to reach the center of the hoop. Time, $t$, equals zero when the ball leaves the player's hands.
2. Determine the velocity, $V_o$, that will result in the ball reaching the center of the hoop at time $t_f$.
3. Create a table consisting of $t$, $x$, $y$ for $0 \le t \le t_f$ in steps of $t_f/10$. Carry variables to 4 decimal places. Print the table to an output file, including $t_f$ and $V_o$.
4. Create a plot of $y$ versus $x$.

Hint: Solve Equation 2.5 for $V_o$ and substitute the expression for $V_o$ into Equation 2.6, giving an expression involving $t$, $x$, and $y$. In that expression, set $t = t_f$, $x = x_f$, and $y = y_f$. In the resulting equation, $t_f$ is the only unknown. Use this expression in your MATLAB program to solve for $t_f$.

**E2.5.** A small sphere moving though a fluid at a slow velocity will have a drag force acting on it, which is described by Stokes' Law. The sphere could be a dust particle or a raindrop moving in air, or a ball bearing moving in oil. The drag force described by Stokes' Law is

$$D = 6\pi R\mu V \qquad (2.7)$$

where:
 $D$ is the drag
 $R$ is the radius of the sphere
 $\mu$ is the viscosity of the fluid
 $V$ is the velocity of sphere

Let us consider a steel ball bearing dropped in oil (see Figure 2.21) with an initial velocity of zero. The ball bearing will drop with a varying velocity until it reaches a final velocity (terminal velocity, $V_T$). The forces acting on the ball bearing are the gravitational force, $W$, buoyancy force, $B$, and the drag force, $D$. The buoyancy force is equal the weight of the fluid displaced. The equations for $W$ and $B$ are

$$W = \rho_{steel}\, \upsilon g \qquad (2.8)$$

$$B = \rho_{oil}\, \upsilon g \qquad (2.9)$$

where:
 $\rho_{steel}$ is the mass density of steel
 $\rho_{oil}$ is the mass density of oil
 $\upsilon$ is the volume of sphere = $4/3\, \pi R^3$
 $g$ is the gravitational constant

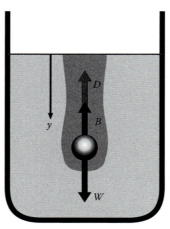

**FIGURE 2.21**
Ball bearing in oil.

# MATLAB® Fundamentals

The terminal velocity occurs when

$$W - B - D = 0 \qquad (2.10)$$

By applying Newton's second law to the sphere we can determine V and $V_T$ that are

$$V_T = \frac{W - B}{6\pi R \mu} \qquad (2.11)$$

$$V = V_T \left( 1 - e^{-\frac{6\pi R \mu g}{W} t} \right) \qquad (2.12)$$

Take $\mu = 3.85$ (N−s)/m², $R = 0.01$ m, $\rho_{steel} = 7910$ kg/m³, $\rho_{oil} = 899$ kg/m³, $g = 9.81$ m/s²

Create a MATLAB program that will

1. Determine the weight of the ball bearing, W.
2. Determine the buoyancy force, B.
3. Determine the terminal velocity, $V_T$.
4. Use a while loop to determine V as a function of time, for $0 \le t \le 0.3$ s in steps of 0.01 s.
5. Create and print to a file values for W, B, and $V_T$ and a table containing t and V.
6. Plot V versus t.

**E2.6.** The voltage in a parallel resistance, inductor, and capacitor (RLC) circuit (see Figure 2.22) is given by

$$v = \exp\left(-\frac{1}{2RC}t\right)\left\{A\exp\left(\sqrt{\left(\frac{1}{2RC}\right)^2 - \frac{1}{LC}}\,t\right) + B\exp\left(\sqrt{\left(\frac{1}{2RC}\right)^2 - \frac{1}{LC}}\,t\right)\right\} \qquad (2.13)$$

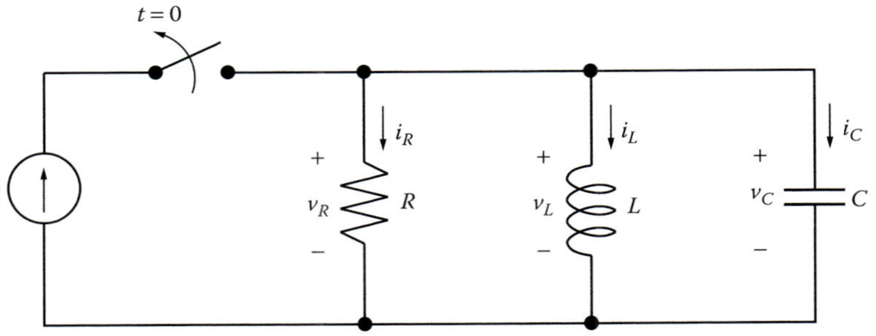

**FIGURE 2.22**
A parallel resistance, inductor, and capacitor (RLC) circuit.

Take R = 10 ohm, L = 1.0e-2 henry, C = 1.0e-6 farad, A = 6.0 V, and B = −8.9 V.

1. Determine $v(t)$ for $0 \leq t \leq 5.0 \times 10^{-4}$ seconds. Use 100 subdivisions on the time domain.
2. Print out a table of $v$ versus $t$ every fourth subdivision.
3. Create a plot of $v$ versus $t$.

## 2.10 Input

Engineers who carry out tests on a piece of equipment may need to enter data into an existing computer program for analysis. There are several commands that may be used to enter data from a data file. These are the `load`, `dlread`, and `fscanf` commands. We will discuss them one at a time. The analysis program may also ask the user to input data from the keyboard. To enter data from the key board, use MATLAB's `input` command, which makes the program interactive.

### 2.10.1 The Load Command

One of the commands that allow the user to enter data from a file is the load command. The data file is likely to only contain numbers and would have a specific name. The syntax for the command is

```
load filename.txt
```

*The input file must have the same number of columns in each row and must be in the same folder as the program loading the data file. The data file should only have numbers.* Suppose we had the following data file named *atm_properties* and we wish to enter the data into a program. Here, the first column is altitude in meters, the second column is temperature in degrees Kelvin (K), the third column is pressure in Pascal (Pa), and the fourth column is density in kilogram/meter³ (kg/m³).

| 0 | 288.15 | 1.0133e+005 | 1.2252 |
| 1000 | 281.65 | 8.9869e+004 | 1.1118 |
| 2000 | 275.15 | 7.9485e+004 | 1.0065 |
| 3000 | 268.65 | 7.0095e+004 | 0.9091 |
| 4000 | 262.15 | 6.1624e+004 | 0.8191 |
| 5000 | 255.65 | 5.4002e+004 | 0.7360 |

## MATLAB® Fundamentals 51

**NOTE:** *Before you can run the following example (Example 2.9), you need to create the data file* shown on the previous page. To do this, copy the data, then go to MATLAB and open up a *new script window* and paste the data into the new *script window*. Then click on the Save icon and save the file as *atm_properties.txt*. To save the file as a **.txt** file, click on the down arrow in the *Save as type* box and select *All files (\*.\*)*.

Alternatively, you can open up a *new script window* and type in the data, and then follow the instructions described above. Try typing or copying the following program into the Editor Window in MATLAB and running it.

**Example 2.7**

```
% Example_2_7.m
% This program uses the load command to load the data in the file
% named atm_properties.txt into this program. Column 1 is the
% altitude, column 2 is the temperature, column 3 is the pressure
% and column 4 is the density. Altitude is in meters (m),
% temperature is in degrees Kelvin (K), pressure is in Pascal (Pa)
% and density is in (kg/m^3).
% The program also is an example of using the colon operator to
% create the vectors z,T,p and rho.
% The program also demonstrates the use of the fprintf command.
% The program prints out elements of the vectors z,T,p and rho.
clear; clc;
load('atm_properties.txt');
% establishing variable names to scanned file.
z=atm_properties(:,1);
T=atm_properties(:,2);
p=atm_properties(:,3);
rho=atm_properties(:,4);
fprintf('  z(m)       T(K)       p(Pa)      rho(m^3/kg)  \n');
fprintf('------------------------------------------------\n');
for j=1:length(z)
    fprintf('%5.0f    %8.2f    %10.3e    %8.4f \n',z(j),T(j),p(j),rho(j));
end
```
------

**Program Results:**

```
  z(m)      T(K)       p(Pa)      rho(m^3/kg)
---------------------------------------------
     0    288.15    1.013e+05    1.2252
  1000    281.65    8.987e+04    1.1118
  2000    275.15    7.949e+04    1.0065
  3000    268.65    7.010e+04    0.9091
  4000    262.15    6.162e+04    0.8191
  5000    255.65    5.400e+04    0.7360
>>
```
------

### 2.10.2 The `dlmread` Command

An alternative to the `load` command is the `dlmread` command. This command will read an ASCII delimited file. All data in the file must be numeric. In this example, the entire data file in *atm_properties.txt* is specified as a matrix Y consisting of six rows and four columns. Then the colon operator is used to create vectors z, T, p, and rho. To demonstrate the use of the `dlmread` command, we will modify Example 2.7 by replacing the lines starting with

```
load (atm_properties.txt);
```

and ending with

```
rho = atm_properties(:,4);
```

with

```
Y=dlmread('atm_properties.txt');
z=Y(:,1);
T=Y(:,2);
p=Y(:,3);
rho=Y(:,4);
```

The modification of Example 2.7 (omitting the comment lines) would be

```
clear; clc;
Y=dlmread('atm_properties.txt');
% establishing variable names to scanned file.
z=Y(:,1);
T=Y(:,2);
p=Y(:,3);
rho=Y(:,4);
fprintf('    z(m)      T(K)        p(Pa)      rho(m^3/kg) \n');
fprintf('-----------------------------------------------------------\n');
for j=1:length(z)
    fprintf('%5.0f %8.2f %10.3e %8.4f \n',z(j),T(j),p(j),rho(j));
end
```

### 2.10.3 `fscanf` Command

Students who have a background in C/C++ may use the `fscanf` command to enter data into a program. The commands necessary to this are shown below.

```
A = zeros(n, m);
fi = fopen('filename.txt','r');
[A] = fscanf(fi,'%f',[n,m]);
```

## MATLAB® Fundamentals

where $n \times m$ is the number of elements in the data file. The `'r'` in the fopen statement tells MATLAB that this file is for reading in data. The $n \times m$ matrix is filled in column order. Thus, rows become columns and columns become rows.

The following example program enters the data in *atm_properties.txt* into the program.

**NOTE:** *Before you can run program Example_2_8.m, the data file* atm_properties.txt *had to be created.*

### Example 2.8

```
% Example_2_8.m
% This program uses fscanf command to load the data in the file
% named atm_propeties.txt into this program. Column 1 is the
% altitude, column 2 is the temperature, column 3 is the pressure
% and column 4 is the density. Altitude is in meters (m)
% temperature is in degrees Kelvin (K), pressure is in Pascal (Pa)
% and density is in (kg/m^3).
clear; clc;
fi = fopen('atm_properties.txt','r');
% Print A to the screen and see that columns of the data file
% became rows.
A = fscanf(fi,'%f',[4,6])
% establishing variable names to scanned file.
z=A(1,:);
T=A(2,:);
p=A(3,:);
rho=A(4,:);
fprintf('z(m)     T(K)       p(Pa)      rho(m^3/kg)  \n');
fprintf('-------------------------------------------\n');
for j=1:length(z)
     fprintf('%5.0f    %8.2f    %10.3e    %8.4f \n',z(j),T(j),p(j),rho(j));
end
----------------------------------------------------------------
```

### Program Results:

```
z(m)    T(K)      p(Pa)     rho(m^3/kg)
-------------------------------------------
   0   288.15   1.013e+05   1.2252
1000   281.65   8.987e+04   1.1118
2000   275.15   7.949e+04   1.0065
3000   268.65   7.010e+04   0.9091
4000   262.15   6.162e+04   0.8191
5000   255.65   5.400e+04   0.7360
>>
----------------------------------------------------------------
```

### 2.10.4 The `input` Command

The MATLAB command that the programmer can use to have the user enter data from the keyboard is the `input` command. The program should pause and move the cursor to the Command Window (without providing a prompt sign) waiting for the user to enter the data requested. However, in MATLAB version R2016A, the cursor stays in the Editor Window. This is a bug in this version of the MATLAB program. This bug was eliminated in MATLAB version R2016B. This was not a problem in earlier versions of MATLAB. The use of the `input` command makes the program interactive. Suppose that you are the programmer and you wish to have the user enter a matrix named z, from the keyboard, use

```
Z = input('Enter matrix Z enclosed by brackets \n')
```

The user will see the following on the screen:

**Enter matrix Z enclosed by brackets**

If you are using MATLAB version R2016A, you will need to click on the Command Window to enter the data. If you are not using MATLAB version R2016A, the user can type information in the Command Window without having to first click in the Command Window. The user should then type in something like

```
[ 5.1 6.3 2.5; 3.1 4.2 1.3 ]
```

Thus, $Z = \begin{bmatrix} 5.1 & 6.3 & 2.5 \\ 3.1 & 4.2 & 1.3 \end{bmatrix}$.

Note that the argument to `input` command is a character string enclosed by the single quotation marks. The character string will be printed to the screen as shown above. If the response to the input statement is a character or a string, you need to enclose the character or the string with single quotation marks. However, you can avoid this requirement by entering a second argument of `'s'` to the `input` command as shown in the following statement:

```
response = input('Print Z to a file? (y/n):\n', 's')
```

In this case, the user can respond with either a **y** or **n** (without single quotation marks). An example using this concept will be given in Chapter 3.

## MATLAB® Fundamentals

**E2.7.** Write a MATLAB program that uses the `input` command to enter the following three vectors:

$$Z = [0\ 1000\ 2000\ 3000\ 4000\ 5000]$$

$$T = [288.1\ 281.6\ 275.1\ 268.6\ 262.1\ 255.6]$$

$$\text{rho} = [1.2252\ 1.1118\ 1.0065\ 0.9091\ 0.8191\ 0.7360]$$

Z is altitude in m, T is temperature in K, and rho is density in kg/m³. Then plot T versus Z and rho versus Z.

### REVIEW 2.5

1. Name four commands that can be used in a script to input data into the workspace.
2. Which of the four commands makes the program interactive?

## 2.11 More on MATLAB Graphics

### 2.11.1 The `figure` Command

As mentioned earlier, if a program involves creating more than one plot, you need to include the statement `figure` after each plot command (except the last), otherwise only the last plot will appear. The following example program produces two separate plots.

**Example 2.9**

```
% Example_2_9.m
% This program creates two separate plots.
% First y1=t^2/10 is plotted with 0 <= t <= 10,
% then y2=t^3/100 is plotted over the same t range.
% To plot y1 and y2 vs. and t, they need to be made vectors.
clear; clc;
t=0:0.5:10;
for n=1:length(t)
    y1(n)=t(n)^2/10;
    y2(n)=t(n)^3/100;
end
plot(t,y1), xlabel('t'), ylabel('y1'), grid, title('y1 vs. t');
figure;
plot(t,y2), xlabel('t'), ylabel('y2'), grid, title('y2 vs. t');
```

**Program Results:**
See Figure 2.23a and b.

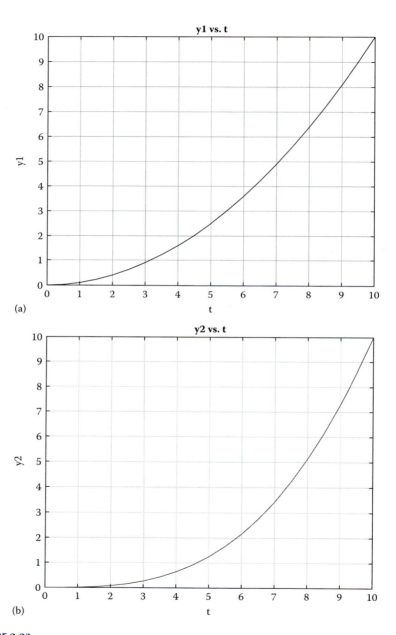

**FIGURE 2.23**
(a) Plot of y1 versus t and (b) plot of y2 versus t.

### 2.11.2 Multiple Plots

Suppose in matrix $A$, shown below, we wished to plot column 2 versus column 1, column 3 versus column 1, and column 4 versus column 1.

$$A = \begin{bmatrix} t_1 & y_1 & z_1 & w_1 \\ t_2 & y_2 & z_2 & w_2 \\ \vdots & \vdots & \vdots & \vdots \\ t_n & y_n & z_n & w_n \end{bmatrix}$$

We could let $T = A(:,1)$, $Y = A(:,2)$, $Z = A(:,3)$, and $W = A(:,4)$, giving

$$T = \begin{bmatrix} t_1 \\ t_2 \\ \vdots \\ t_n \end{bmatrix}, Y = \begin{bmatrix} y_1 \\ y_2 \\ \vdots \\ y_n \end{bmatrix}, Z = \begin{bmatrix} z_1 \\ z_2 \\ \vdots \\ z_n \end{bmatrix}, W = \begin{bmatrix} w_1 \\ w_2 \\ \vdots \\ w_n \end{bmatrix}$$

Then to plot $Y$ versus $T$, $Z$ versus $T$ and $W$ versus $T$ all on the same graph, we would write,

```
plot(T,Y,T,Z,T,W);
```

Of course, we could have avoided the additional steps by writing

```
plot(A(:,1),A(:,2),A(:,1),A(:,3),A(:,1),A(:,4))
```

To identify which curve goes with which variable, you can add text to the plot with the command,

```
text(x,y,'text statement');
```

where (x, y) are the coordinates on the graph where the text statement will start.

Multiple curves on the same graph can be distinguished by color coding the curves.

Available color types:

| | |
|---|---|
| black | 'k' |
| blue | 'b' |
| green | 'g' |
| red | 'r' |
| cyan | 'c' |
| yellow | 'y' |

Multiple curves on the same graph can also be distinguished by using different types of lines.

Available line types:

| solid | (default) |
|---|---|
| dashed | '--' |
| dashed-dot | '-.' |
| dotted | ':' |

Alternatively, you can create a marker plot of discrete points (without a line) by using one of these marker styles:

| point | '.' |
|---|---|
| plus | '+' |
| star | '*' |
| circle | 'o' |
| x-mark | 'x' |
| diamond | 'd' |

The `legend` command may also be used in place of the text command to identify the curves. The format for the legend command is

```
legend('text1', 'text2')
```

The `legend` box may be moved by clicking on the box and dragging it to the desired position.

You can also change the axis limits in a plot by using the command

```
axis([xmin xmax ymin ymax])
```

(See Example 2.11)

### Example 2.10

The following example illustrates a multiple plot program:

```
% Example_2_10.m
% This program creates a simple table and a multiple plot.
% First a table of y1=t^2/10 and y2=t^3/100 is created.
% To plot y1, y2 vs. and t, they need to be made vectors.
% y1 and y2 vs. t are plotted on the same graph.
clear; clc;
t=0:10;
for n=1:length(t)
    y1(n)=t(n)^2/10;
    y2(n)=t(n)^3/100;
end
```

## MATLAB® Fundamentals

```
% By making t, y1 and y2 as vectors, their values can be printed
% outside the for loop that created them.
% Column headings
fprintf('    t             y1            y2   \n');
fprintf('---------------------------------------\n');
for n=1:length(t)
    fprintf('%8.1f       %10.2f      %10.2f \n',t(n),y1(n),y2(n));
end
% Create the plot, y1 as a solid line, y2 as a dashed line.
% Note: the variables t, y1,y2 need to be vectors in the plot
% command.
plot(t,y1,t,y2,'--');
xlabel('t'), ylabel('y1,y2'), grid, title('y1 and y2 vs. t');
% Plot identification is also established by adding text to the plot.
text(6.5,2.5,'y2');
% In the above statement, 6.5 is the abscissa position and 2.5 is
% the ordinate position where the 'y1' label will be positioned.
text(4.2,2.4,'y1'),
% We can also use the legend command to identify the curves
legend('y1','y2');
```
---

**Program Results:**

```
    t             y1            y2
---------------------------------------
   0.0           0.0000        0.0000
   1.0           0.1000        0.0100
   2.0           0.4000        0.0800
   3.0           0.9000        0.2700
   4.0           1.6000        0.6400
   5.0           2.5000        1.2500
   6.0           3.6000        2.1600
   7.0           4.9000        3.4300
   8.0           6.4000        5.1200
   9.0           8.1000        7.2900
  10.0          10.0000       10.0000
>>
```

See Figure 2.24.
---

### 2.11.3 The hold on Command

In the above example, we used a single plot command to plot both y1 and y2, that is, plot(t,y1,t,y2,'--'). However, we could also have plotted both y1 and y2 on the same graph by plotting each separately with the command hold on between the plots. We would be superimposing the second plot onto the first plot. To do this, replace the plot command plot(t,y1,t,y2,'--') with

```
plot(t,y1);
hold on
plot(t,y2);
xlabel('t'), ylabel('y1,y2'), grid, title('y1 and y2 vs. t');
```

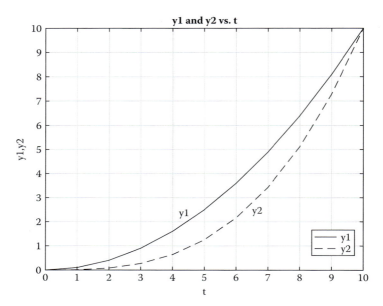

**FIGURE 2.24**
Plots of y1 and y2 on the same graph.

### Example 2.11

The following example illustrates the plotting of trigonometric functions. The example also illustrates that simple arithmetic can be used in the arguments of the trigonometric functions. This is also true for other built-in MATLAB functions. If the resulting curves are not smooth than we would need to use more points to properly display the curves.

```
% Example_2_11.m
% This script calculates both sin(2x/3), sin(2x/3)^2
% and cos(2x/3+pi) for -pi <= x <= pi. The x domain is subdivided
% into 50 subdivisions. The script plots the 3 functions and
% determines the absolute maximum values of the vectors fsin, fsinsq
% and fcos and prints those values to the screen.
clear; clc;
x=-pi:2*pi/50:pi;
for i=1:length(x)
    fsin(i)=sin(2*x(i)/3);
    fsinsq(i)=sin(2*x(i)/3)^2;
    fcos(i)=cos(2*x(i)/3+pi);
end
```

# MATLAB® Fundamentals

```
fsin_max=max(abs(fsin)); fcos_max=max(abs(fcos));
fsinsq_max=max(fsinsq);
fprintf('fsin_max=%10.5f, fcos_max=%10.5f \n',fsin_max, fcos_max);
fprintf('fsinsq_max=%10.5f \n',fsinsq_max);
plot(x, fsin, x,fcos,'--',x, fsinsq,'-.'), xlabel('x'),
ylabel('fsin, fcos, fsinsq'), grid,
title('fsin, fcos, fsinsq vs. x'), legend('fsin','fcos','fsi nsq');
```

**Program Results:**

From the Command Window:

```
fsin_max=    0.99978, fcos_max= 1.00000
fsinsq_max= 0.99956
>>
```

See Figure 2.25.

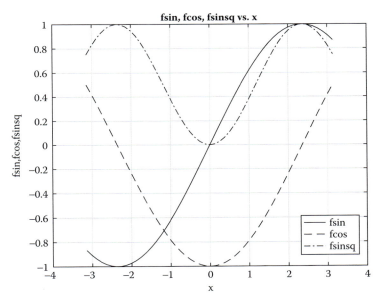

**FIGURE 2.25**
Plot of fsin, fcos, and fsinsq versus x.

### 2.11.4 Plotyy Command

Suppose we have two functions of the same variable but the numerical range of the two functions differ significantly and we would like to display the functions on the same plot. This can be done using the plotyy function. In the next example we plot position, y, and velocity, V of a free falling body vs. time on the same graph (see Example 2.6).

**Example 2.12**

```
% Example_2_12.m
% This script is a modification of Example 2.5. In this script
% both y and V axes are plotted on the same graph. The y axis
% is on the left side and the V axis is on the right side.
clear; clc;
Vo=10.0; g=9.81; t=0:0.1:2;
for i=1:length(t)
    V(i)=Vo-g*t(i); y(i)=Vo*t(i)-0.5*g*t(i)^2;
end
plotyy(t,y,t,V), xlabel('t(s)'), title('y vs. t and V vs. t'), grid,
yyaxis left, axis([0 2 0 10]), ylabel('y(m)'), text(0.32,2.5,'y');
yyaxis right; axis([0 2 -10 10]), ylabel(' V(m/s)'), text(0.6,5.0,'V');
```

**Program Results:**

See Figure 2.26.

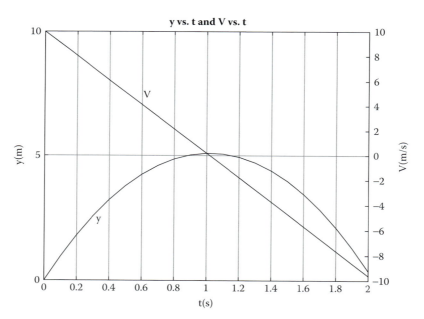

**FIGURE 2.26**
Plot of y versus t and V versus t on the same graph.

## 2.11.5 The subplot Command

Suppose you want to plot each of several curves as a separate plot, but all on the same page. The `subplot` command provides the means to do so. The command `subplot(m,n,p)` breaks the page into an m by n matrix of small plots, and p selects the matrix position of the plot. The `subplot` command is a positioning command and not a plot command. The following example demonstrates the use of the `subplot` command.

**Example 2.13**

```
% Example_2_13.m
% This program is an example of the use of the subplot command.
% Values of y1, y2, y3 and y4 are constructed as
% vectors. Separate plots of y1 vs. t, y2 vs. t, y3 vs. t,
% and y4 vs. t are plotted on the same page.
clc; clear;
t=0:0.5:10;
for n=1:length(t)
    y1(n)=t(n)^2/10;
    y2(n)=sin(pi*t(n)/10);
    y3(n)=exp(t(n)/2);
    y4(n)=sqrt(t(n));
end
subplot(2,2,1),
plot(t,y1), grid, title('y1 vs. t'), xlabel('t''), ylabel('y1');
subplot(2,2,2),
plot(t,y2), grid, title('y2 vs. t'), xlabel('t'), ylabel('y2');
subplot(2,2,3),
plot(t,y3), grid, title('y3 vs. t'), xlabel('t'), ylabel('y3');
subplot(2,2,4),
plot(t,y4), grid, title('y4 vs. t'), xlabel('t'), ylabel('y4');
```
------------------------------------------------------------------------

**Program Results:**
See Figure 2.27.
------------------------------------------------------------------------

## 2.11.6 Bar Charts

I find that bar charts appear more often in business related topics than in engineering topics. For example, you might wish to compare several companies yearly profit percentages. A convenient way to do this is by the use of a bar chart. The syntax the bar charts are

```
bar(y)
bar(x,y)
bar(___, width)
bar(___, style)
bar(___, color)
```

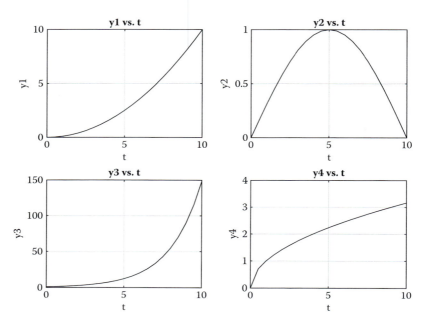

**FIGURE 2.27**
Plots of y1, y2, y3, and y4 versus t on the same page.

See MATLAB help for more examples on bar charts.

**Example 2.14**

In this example, we compare the profit percentage gained for year 2015 for Companies A, B, C, D, E, F, and G. These are given in Table 2.1

```
% Example_2_14.m
% This script is an example of creating a bar chart.
% The script lists and plots the percentage gains in profits for
% several companies for the year 2015.
clear; clc;
y = [2.51 -0.13 3.16 4.72 1.2 6.5 3.8];
bar(y,0.4);
ylabel('% profit, year 2015');
title('1=CO.A, 2=CO.B, 3=CO.C, 4=CO.D, 5=CO.E, 6=CO.F, 7=CO.G');
```
---

**Program Results:**

See Figure 2.28.
---

## MATLAB® Fundamentals

**TABLE 2.1**

Companies Profits Percentage Gain for 2015

| Company | % Profit Gain for Year 2015 |
|---|---|
| A | 2.51 |
| B | −0.13 |
| C | 3.16 |
| D | 4.72 |
| E | 1.60 |
| F | 6.50 |
| G | 3.80 |

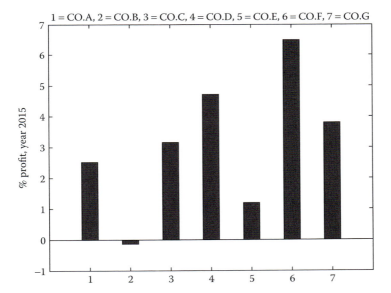

**FIGURE 2.28**
Bar chart. Percent profit gain for several companies for year 2015.

### 2.11.7 Pie Charts

As with bar charts, pie charts appear more often in business and finances than in engineering. For example, you might wish to know what percentage of your investments are in U.S. stocks, foreign stocks, mutual funds, bonds, and cash.

The syntax for the pie chart for this example could be

```
X = 1:5;
labels = {'US stocks','foreign stocks','mutual funds',...
'bonds','cash'};
pie(X,labels)
```

## Example 2.15

Suppose we wished to plot the percentage of several different types of investments made by a particular investor. Table 2.2 gives the percentage of different types of investments for that individual.
The program follows:

```
% Example_2_15.m
% This program gives the percentage of various types of investments
% made by a particular individual. The percentages are displayed in
% a pie chart.
clear; clc;
x=[43 12 15 20 10];
labels = {'US stocks','foreign stocks','mutual funds',...
'bonds','cash'};
pie(x, labels);
```
---
**Program Results:**

See Figure 2.29.
---

### TABLE 2.2

Investment Percentages

| Investments | Percentage |
|---|---|
| US stocks | 43 |
| Foreign stocks | 12 |
| Mutual funds | 15 |
| Bonds | 20 |
| Money market | 10 |

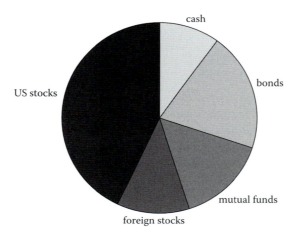

**FIGURE 2.29**
Pie chart for several different types of investments.

## Example 2.16

- The `sprintf` Command

The `sprintf` is the same as `fprintf` command except that it returns the print data as a MATLAB variable rather than writing to the Command Window or to a file. When plotting several different plots on the same page, you may wish to vary the titles of the plots depending on the specific variable defined in the program. This is demonstrated in the following example:

```
% Example_2_16.m
% This program is an example of the use of the subplot and
% the sprintf commands.
% Plots of y=sin(k*pi*t/L) for several values of k are created and
% plotted on the same page.
clc; clear;
t=0:0.1:10;
k=[2 4 6 8];
L=10;
for m=1:length(k)
    for n=1:length(t)
        y(n)=sin(k(m)*pi*t(n)/L);
    end
    subplot(2,2,m), plot(t,y), xlabel('t'), ylabel('y'), grid,
    title(sprintf('y vs.t, k=%3i \n',m));
end
```

**Program Results:**

See Figure 2.30.

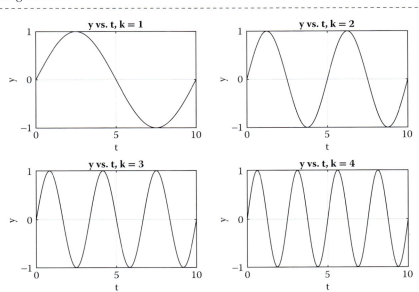

**FIGURE 2.30**
Four plots of y versus t, all on one page. Each plot is for a different value of k.

**Exercises**

**E2.8.** This exercise involves plotting the temperatures of a spherical object dropped into a fluid contained in a vertical circular cylinder. The temperature variation of both the sphere, $T\_sphere$, and the fluid, $T\_fluid$, are given as a function of time, $t$, in the following three vectors:

$$t = [0\ 0.04\ 0.08\ 0.12\ 0.16\ 0.20\ 0.24\ 0.28\ 0.32\ 0.36\ 0.40\ 0.44\ 0.48\ 0.52\ 0.56\ 0.60]$$

$$T\_sphere = [150\ 124\ 104\ 89\ 77\ 67\ 60\ 54\ 49\ 46\ 43\ 41\ 39\ 38\ 37\ 36]$$

$$T\_fluid = [20.0\ 22.9\ 25.2\ 27.0\ 28.3\ 29.5\ 30.3\ 31.0\ 31.9\ 32.2\ 32.5\ 32.7\ 32.8\ 32.9]$$

In MATLAB, create a plot of both $T\_sphere$ and $T\_fluid$ versus $t$ on the same graph, $t$ is in seconds and $T\_sphere$ and $T\_fluid$ are in degrees C.

**E2.9.** This exercise involves the $x$ position and $x$ component of the velocity, $u$, of a package dropped from an airplane as a function of time, $t$. These variables are specified in vectors $x$ and $u$ and $t$, respectively.

$$t = [0.0\ 0.5\ 1.0\ 1.5\ 2.0\ 2.5\ 3.0\ 3.5\ 4.0\ 4.5\ 5.0]$$

$$x = [0.0\ 23.6\ 44.7\ 63.6\ 80.6\ 95.8\ 109.5\ 121.7\ 132.7\ 142.4\ 150.9]$$

$$u = [50.0\ 44.6\ 39.9\ 35.8\ 32.2\ 28.9\ 25.9\ 23.1\ 20.6\ 18.2\ 15.9]$$

Create a MATLAB program that will plot $x$ versus $t$ and $u$ versus $t$ as two separate plots, but both on the same page.

### 2.11.8 Greek Letters and Mathematical Symbols

Greek letters and mathematical symbols can be used in `xlabel`, `ylabel`, `title`, and `text` by spelling out the Greek letter and preceding it with a '\' (backslash character). Thus, to display ω, use \omega, and to display β, use \beta.

**Example:**
```
ylabel('\omega'), title('\omega vs. \beta'), text(10,5,'\omega');
```

For an additional list of Greek symbols and some special characters, see Appendix A. You may also occasionally need to print a "'" character in your label or title. In this case, use a double-quotation mark as shown here '' to *escape* the single-quote character in your string. Thus, to generate the plot title "Signal 'A' vs. Signal 'B'", you would type

```
title('Signal ''A'' vs. Signal ''B''')
```

## MATLAB® Fundamentals

### 2.11.9 Interactively Annotating Plots

As an alternative to adding the `xlabel`, `ylabel`, and `title` commands into your program, you can create the plot, then click on the `Insert` option in the menu bar in the plot window and choose `X Label` from the dropdown menu. This will highlight a box in which you can type in the abscissa variable name. You can repeat this process for the `Y Label` and the `Title` of the plot. Other options available in the `Insert` Menu are `TextBox`, `Text Arrow`, `Arrow`, and others. When you click any one of these options, a cross-hair will appear and you can then move the item to the location where you want it to appear, then left-click the mouse to fix the location. You can then type in the desired text. To remove the outlines of a `TextBox`, place the cursor in the `TextBox` and right-click the mouse. This will bring up a dropdown menu, then select `Line Style`, and then left-click on `none`. This will remove the lines from the `TextBox`.

### 2.11.10 Saving Plots

To save a plot, click on the `File` in the plot window and select the `Save` option from the dropdown menu. This produces a window where you can enter a file name. The disadvantage of this method is that if you decide to rerun the script, the items that you **manually** inserted will not be saved. If you wish to copy the figure into a report, you can click on `Edit` in the plot window, and then select `Copy Figure` from the dropdown menu. You can then paste the figure into your report. If you need a monochrome version of your plot (for best reproduction on a photocopier), you can make all of your curves black by choosing `File` from the task bar menu, then selecting `Export Setup` from the dropdown menu. This will open a window in which you need to click on `Rendering`, and change the `Colorspace` to `black and white`.

There are many more options available in the plot window, however we leave it to the student to explore it further.

---

### REVIEW 2.6

1. When there is more than one function plotted on a graph, what are the ways to identify which curve goes with which function?
2. What is the name of the function that will allow you to plot several graphs on one page?
3. How does one enter Greek symbols into a plot?
4. What are the commands that will allow you to enter text onto a plot once the plot has been created?

## Projects

**P2.1.** A tennis player on serve wishes to place the tennis ball close to the outside line of the service box when the ball hits the ground (see Figure P2.1a and b). The horizontal distance from the point where the ball leaves the racket to where the ball hits the ground is 19.33 m. The vertical distance, $y_o$, above the ground when the ball leaves the racket is 2.36 m. The angle that the ball makes with the horizontal on leaving the players racket is 5.7° pointing down. Neglecting drag, the governing equations describing the motion of the ball are

$$x = V_o \cos(\vartheta) t \tag{P2.1a}$$

$$y = -\frac{g}{2} t^2 - V_o \sin(\vartheta) t + y_o \tag{P2.1b}$$

In the above equations, $x$ and $y$ are in (m), $t$ is in (s) and $V_o$ is in (m/s).

Let $(x_f, y_f)$ be the $x$ and $y$ positions where the ball hits the ground and $t_f$ the time when this occurs.

1. Determine the time, $t_f$. Time, $t$, equals zero when the ball leaves the racket.
2. Determine the velocity, $V_o$, that will result in the ball reaching the ground at time $t_f$.

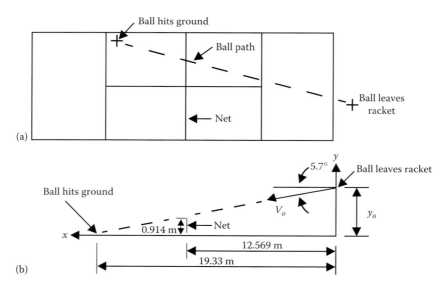

**FIGURE P2.1**
Tennis player on serve: (a) plan view and (b) vertical view.

3. Create a table consisting of t, x, y for $0 \le t \le t_f$ in steps of $t_f/10$. Carry variables to 4 decimal places. Print the table to an output file, including $t_f$ and $V_o$.
4. Create a plot of y versus x.
5. Determine the height of the tennis ball at the position of the net.

**P2.2.** A batter in a baseball game hits a ball to right center field. The ball leaves the bat at a 30° angle with the horizontal at a speed of $V_o$ and at a height of 1.5 m above the ground. The center fielder is 71 m from home plate and the angle that the horizontal line connecting the center fielder with home plate makes angle of 10° with the horizontal path of the ball, see Figure P2.2a and b. The center fielder sees the direction of the fly ball and starts to run toward the path of the ball at an average speed, S and 0.5 s after the ball is hit. The center fielder catches the ball when it is just 1.8 m above the ground and 91 horizontal meters from the initial position of the ball as it leaves the bat. Neglecting drag, the governing equations describing the motion of the ball are

$$x = V_o \cos(\vartheta) t \tag{P2.2a}$$

$$y = -\frac{g}{2} t^2 - V_o \sin(\vartheta) t + y_o \tag{P2.2b}$$

1. What is the time of flight, $t_f$, when the ball is caught?
2. What is the initial velocity, $V_o$, of the ball when it leaves the bat?

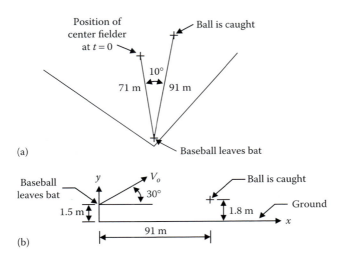

**FIGURE P2.2**
Batter hitting baseball: (a) Plan view and (b) vertical view.

3. What is the average speed, S, of the center fielder as a runs to catch the ball.
4. Create a table consisting of t, x, y for $0 \le t \le t_f$ in steps of $t_f/10$. Carry variables to 4 decimal places. Print the table to the Command Window, include $t_f$, $V_o$, and S.
5. Create a plot of y versus x.

**P2.3.** Although atmospheric conditions vary from day-to-day, it is convenient for design purposes, to have a model for atmospheric properties as a function of altitude. The U.S. Standard Atmosphere, modified in 1976, is such a model. For altitudes less than or equal to 11,000 m, the governing equations for the air temperature, pressure, and density are as follows:

$$p = p_o \left(1 - \frac{\lambda z}{T_o}\right)^{\frac{g}{\lambda R}} \tag{P2.3a}$$

$$T = T_o - \lambda z \tag{P2.3b}$$

$$\rho = \frac{p}{RT} \tag{P2.3c}$$

where:
z is the altitude
$T_o$ = 288.15 K (the temperature at z = 0)
$p_o$ = 1.01325 × 10⁵ Pa (the pressure at z = 0)
R = 287 J/(kg-K) (the gas constant for air)
g = 9.81 m/s² (the gravitational constant for air)
$\lambda$ = 0.0065 K/m (the lapse rate)
$\rho$ is the air density (kg/m³)

Calculate atmospheric properties of temperature, T, pressure, p, and density, ρ, every 1000 m from z = 0 (sea level) to z = 11,000 m and print the results to a file in a table format. Also plot T versus z, p versus Z, and ρ versus z as three separate plots, all on the same page.

**P2.4.** The properties of specific volume, v, and pressure, p, as a function of temperature, T, for carbon dioxide based on the Redlich–Kwong Equation of state are given in Table P2.1:
Plot v versus T and p versus T as two separate plots.

## TABLE P2.1
Specific Volume and Pressure versus Temperature

| T (K) | v (m³/kmol) | p (bar) |
|---|---|---|
| 350 | 0.28 | 7.65 |
| 400 | 0.32 | 8.57 |
| 450 | 0.36 | 9.16 |
| 500 | 0.40 | 9.55 |
| 550 | 0.44 | 9.81 |
| 600 | 0.48 | 10.00 |
| 650 | 0.52 | 10.14 |
| 700 | 0.56 | 10.24 |
| 750 | 0.60 | 10.31 |

**P2.5.** In this project, we consider two cars on a collision course (see Figure P2.3). Each car's initial position and the angle its path makes with the $x$-axis is specified below. The speed of $car_1$ is also specified.

Initial position of $car_1$: $x_1 = 500$ m, $y_1 = 100$ m, and $V_1 = 40$ m/s. $Car_1$ moves in a straight line that makes an angle of 60° with the $x$-axis.

Initial position of $car_2$: $x_2 = 2000$ m, $y_2 = 200$ m. $Car_2$ moves in a straight line and makes an angle of 45° with the $(-x)$ axis.

The collision coordinates are $(x_c, y_c)$. See Figure P2.3.

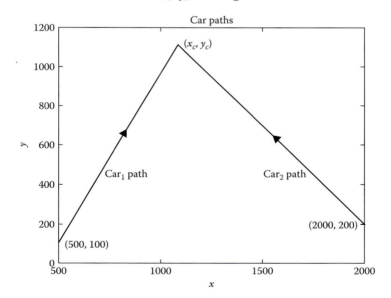

**FIGURE P2.3**
Two cars on a collision path.

We can determine the coordinates of the collision point by writing the equation for the tangent of each line, solving each equation for $y_c$, equating the two $y_c$ expressions, then solving for $x_c$, as shown below.

$$\frac{y_c - y_1}{x_c - x_1} = \tan(60°), \quad \frac{y_c - y_2}{x_2 - x_c} = \tan(45°) \tag{P2.5}$$

Solving each equation for $y_c$ gives

$$y_c = y_1 + (x_c - x_1) \times \tan(60°) \tag{P2.5}$$

$$y_c = y_2 + (x_2 - x_c) \times \tan(45°) \tag{P2.5}$$

Equating the two $y_c$ values gives

$$y_1 + (x_c - x_1) \times \tan(60°) = y_2 + (x_2 - x_c) \times \tan(45°) \tag{P2.5}$$

Solving for $x_c$ gives

$$x_c = \frac{y_2 - y_1 + x_1 \tan(60°) + x_2 \tan(45°)}{\tan(60°) + \tan(45°)} \tag{P2.5}$$

The distance travelled by each car from the initial state to the collision state is

$$d_1 = \sqrt{(x_c - x_1)^2 + (y_c - y_1)^2} = V_1 t_c \tag{P2.5}$$

$$d_2 = \sqrt{(x_2 - x_c)^2 + (y_c - y_2)^2} = V_2 t_c \tag{P2.5}$$

where $t_c$ = time of collision. Equating the $t_c$ from both the above equations give

$$\frac{d_1}{V_1} = \frac{d_2}{V_2} \rightarrow V_2 = V_1 \frac{d_2}{d_1} \tag{P2.5}$$

On line 1:

$$x(t) = x_1 + V_1 \cos(60°) t, \quad y(t) = y_1 + V_1 \sin(60°) t \tag{P2.5}$$

On line 2:

$$x(t) = x_2 - V_2 \cos(45°) t, \quad y(t) = y_2 + V_2 \sin(45°) t \tag{P2.5}$$

Create a MATLAB program that will do the following.

1. Create a plot of the intersecting lines of lengths $d_1$ and $d_2$.

NOTE: You only need to specify two points on the line to plot the line.

2. Determine $V_2$ that will cause the collision to take place.

# MATLAB® Fundamentals

3. Take $t = 0: t_c/5: t_c$ and plot the two lines and the two car's positions at $t_i$, shown as small circles, all on the same graph.

**P2.6.** A formula describing the fluid level, $h(t)$, in a tank as the fluid discharges through a small circular orifice (see Figure P2.4) is

$$\sqrt{h} = \sqrt{h_o} - \frac{C_d A_o}{2 A_T} \sqrt{2g}\, t \quad \quad (P2.6)$$

where:
$C_d$ is the discharge coefficient
$h_o$ is the fluid level in the tank at time, $t = 0$
$A_o$ is the circular area of the orifice having diameter $d$
$A_T$ is the circular cross-sectional area of the tank having diameter $D$

Create a MATLAB program that will

1. Determine vectors $h$ versus $t$, for $0 \le t \le 80$ s.
2. Create a table containing 20 values of $t$ and $h$ (every fourth time step) and print the table to a file and print the file.
3. Create a plot of $h$ versus $t$ and print the plot.

Use the following parameters:
$h_o = 0.3$ m, the tank diameter, $D = 0.8$ m, the orifice diameter, $d = 0.05$ m, $g = 9.81$ m/s² and $C_d = 0.7$.

**P2.7.** When a fluid flows through a pipe there is a pressure drop that is proportional to the pipes length (see Figure P2.5). For a pipe having a circular cross section, the pressure drop, $p_1 - p_2$ [1] is given by

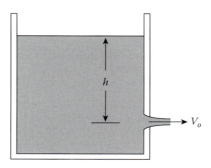

**FIGURE P2.4**
Fluid discharging through a small orifice.

**FIGURE P2.5**
Fluid flow through a pipe.

$$p_1 - p_2 = \frac{\rho V^2}{2} \frac{L}{D} f \qquad (P2.7)$$

where:
$\rho$ is the fluid density (kg/m³)
V is the average fluid velocity in the pipe (m/s)
D is the pipe diameter (m)
L is the pipe length between points 1 and 2 (m)
f is the friction factor

The friction factor has been determined by experiment. For smooth pipes a formula that approximates the experimental data is [5]

$$f = (1.82 \log_{10} \text{Re} - 1.64)^{-2} \qquad (P2.7)$$

where:

$$\text{Re} = \frac{\rho V D}{\mu} \text{ (Reynolds number)} \qquad (P2.7)$$

and $\mu$ = Absolute fluid viscosity (N−s/m²).

Develop a MATLAB program that will calculate

1. f versus Re.
2. V versus Re.
3. $p_1 - p_2$ versus Re.
4. Plot log(Re) on the x-axis and f on the y-axis (semilog plot). Take

Re = [5.0e3 7.5e3 1.0e4 2.5e4 5.0e4 7.5e4 1.0e5 2.5e5 5.0e5 7.5e5 1.0e6 2.5e6 5.0e6 ].
Take $\rho$ = 1000 kg/m³, L = 50 m, D = 0.15 m, and $\mu = 1.52 \times 10^{-3}$ N−s/m²

**P2.8.** The positioning of a piston in an internal combustion engine is shown in Figure 2.18a and b. The piston's position, s, as seen from the crank shaft center can be determined by the Law of cosines, that is,

$$b^2 = s^2 + r^2 - 2sr\cos\theta \qquad (P2.8a)$$

# MATLAB® Fundamentals

or

$$s^2 - (2r\cos\theta)s + (r^2 - b^2) = 0 \tag{P2.8b}$$

where:
 $b$ is the length of the piston rod
 $r$ is the radius of the crankshaft

Equation P2.8b is a quadratic equation in $s$ and therefore

$$s = \frac{1}{2}\left(2r\cos\theta + \sqrt{4r^2\cos^2\theta - 4(r^2 - b^2)}\right) = r\cos\theta + \sqrt{r^2(\cos^2\theta - 1) + b^2}$$

or

$$s = r\cos\theta + \sqrt{b^2 - r^2\sin^2\theta} \tag{P2.8c}$$

The piston is constrained to move in the vertical direction and its position, $s$, varies as the crankshaft rotates. The angle, $\theta$, varies with time, $t$, and can be expressed in terms of the rotational speed, $\omega$, of the crankshaft. The angle v is thus given by

$$\theta = 2\pi\omega t \tag{P2.8d}$$

where $\omega$ is in revolutions per second. Substituting Equation P2.8d into Equation P2.8c gives

$$s(t) = r\cos(2\pi\omega t) + \sqrt{b^2 - r^2\sin^2(2\pi\omega t)} \tag{P2.8e}$$

The piston velocity, $V$, can be obtained by taking the derivative of Equation P2.8e with respect to time giving

$$V(t) = -2\pi\omega r\sin(2\pi\omega t) - \frac{2\pi\omega r^2\sin(2\pi\omega t)\cos(2\pi\omega t)}{\sqrt{b^2 - r^2\sin^2(2\pi\omega t)}} \tag{P2.8f}$$

1. In MATLAB, create a matrix consisting of $s$ versus $t$ and $V$ versus $t$, for $0 \le t \le 0.02$ seconds. Use 50 subdivisions on the $t$ domain. Take $r = 9$ cm, $\omega = 100$ revolutions per second, and $b = 14$ cm. Plot $s$ versus $t$ and $V$ versus $t$ as two separate plots.

2. Using MATLAB's `max` function and the matrix obtained in part (1), determine the approximate maximum velocity and print out those values to the screen.

3. Plot on a single page $s$ versus $t$ for $\omega = [50\ 100\ 150\ 200]$ revolutions per second.

**P2.9.** This project involves plotting the oscillatory motion of a mass in a mass-spring-dashpot system (see Figure 2.19). The governing equation for the position, $y$, of the mass measured from the equilibrium position depends on the values of the spring constant, $k$, the damping factor, $c$, and the mass, $m$.

If, $k/m > (c/2m)^2$, then the mass motion will be damped oscillations and the governing equation describing the motion is

$$y = \exp\left(-\frac{c}{2m}t\right)\left\{A\cos\left(\sqrt{\frac{k}{m}-\left(\frac{c}{2m}\right)^2}\,t\right) + B\sin\left(\sqrt{\frac{k}{m}-\left(\frac{c}{2m}\right)^2}\,t\right)\right\} \quad \text{(P2.9)}$$

The coefficients A and B are determined by initial conditions, which is beyond the scope of this textbook. Given the following parameters:

$$m = 25 \text{ kg}, \; k = 200\frac{\text{N}}{\text{m}}, \; c = 5\frac{\text{N-s}}{\text{m}}, \; A = 0.5 \text{ m}, \; B = \frac{c}{2m} \times \frac{A}{\sqrt{\frac{k}{m}-\left(\frac{c}{2m}\right)^2}}$$

For a complete derivation of Equation P2.9 see Project P2.5 in [3,4].
Determine $y(t)$ for $0 \le t \le 20$ seconds in steps of 0.1 seconds.
The envelope of the solution graph for this case is given by

$$y_{env} = \pm A\exp\left(-\frac{c}{2m}t\right)$$

Plot $y$ versus $t$ and $y_{env}$ versus $t$ on the same graph.

**P2.10.** In this project we consider the voltage, $v$, of a parallel RLC circuit when at $t = 0$, the switch is opened. See Figure 2.22. The governing equation for $v$, depends on the values of $R$, $L$, and $C$.

If $(1/2RC)^2 < 1/LC$, then the solutions are decaying sinusoids over time (*underdamped*) and the governing equation for $v$ is

$$v = \exp\left(-\frac{1}{2RC}t\right)\left\{A\cos\left(\sqrt{\frac{1}{LC}-\left(\frac{1}{2RC}\right)^2}\,t\right) + B\sin\left(\sqrt{\frac{1}{LC}-\left(\frac{1}{2RC}\right)^2}\,t\right)\right\} \quad \text{(P2.10)}$$

For a complete derivation of Equation P2.10 see Project P2.7 in [3,5]. The coefficients A and B are to be determined by initial conditions, which are beyond the scope of this book.

Create a MATLAB program that will calculate and plot $v(t)$ for $0 \leq t \leq 500$ µsec in steps of 5 µsec using the following parameters:

$R = 100\ \Omega$, $L = 1$ mH, $C = 1$ µF, $A = 6.0000$ V, and $B = -0.9608$ V

**P2.11.** This project involves determining the rate that heat, $q$, which is transferred into a house per unit surface area from a section of the exterior walls shown in Figure P2.6. The wall consists of plaster board, insulation, wood sheathing, and brick.

The governing equation describing the rate that heat is transferred, $q$, into a house from the wall section [6] shown in Figure P2.6 is

$$q = \frac{T_o - T_i}{\dfrac{1}{h_o} + \dfrac{L_1}{k_1} + \dfrac{L_2}{k_2} + \dfrac{L_3}{k_3} + \dfrac{L_4}{k_4} + \dfrac{1}{h_i}} \tag{P2.11}$$

where:
$h_o$ is the outside convective heat transfer coefficient
$h_i$ is the inside convective heat transfer coefficient
$k_1$ is the thermal conductivity of brick
$k_2$ is the thermal conductivity of wood sheathing
$k_3$ is the thermal conductivity of insulation
$k_4$ is the thermal conductivity of plaster board

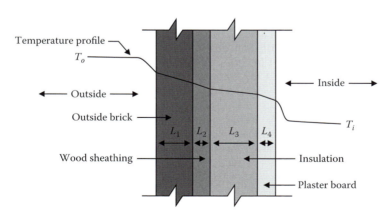

**FIGURE P2.6**
Heat flow through exterior wall.

Create a MATLAB program that determines the rate that heat flows into a house per unit surface area due to the section of the wall described above. Take $T_o$ to vary from 25°C to 40°C in steps of 1.0°C and $T_i = 20°C$. Create a table containing $T_o$ and $q$, include table headings with units. Also, plot $q$ versus $T_o$.

The thickness, $L$, of each material follows:

$$L_4 = 1.3 \text{ cm}, \quad L_3 = 10 \text{ cm}, \quad L_2 = 1.3 \text{ cm} \text{ and } L_1 = 0.7 \text{ cm}$$

The thermal conductivity, $k$, of each of the materials follows:

$$k_4 = 0.48 \frac{W}{m-C}, \quad k_3 = 0.05 \frac{W}{m-C}, \quad k_2 = 0.11 \frac{W}{m-C}$$

and

$$k_1 = 0.69 \frac{W}{m-C}, \quad h_o = 56.8 \frac{W}{m^2-C} \text{ and } h_i = 11.4 \frac{W}{m^2-C}$$

For the interior temperature to remain constant, an air-conditioning system must remove heat at the rate that heat enters the house from the outside as well as any heat that is generated in the interior.

# References

1. Bober, W., Kenyon, R.A., *Fluid Mechanics*, John Wiley & Sons, New York, 1980.
2. Bober, W., The use of the Swamee-Jain formula in pipe network problems, *Journal of Pipelines*, 4, 315–317, 1984.
3. Bober, W., *Introduction to Numerical and Analytical Methods with MATLAB for Engineers and Scientist*, CRC Press, Boca Raton, FL, 2014.
4. Thomson, W. T., *Theory of Vibration with Applications*, Prentice Hall, Englewood Cliffs, NJ, 1972.
5. Bober, W., Stevens, A., *Numerical and Analytical Methods with MATLAB for Electrical Engineers*, CRC Press, Boca Raton, FL, 2012.
6. Holman, J.P., Heat Transfer, 9th Ed., McGraw-Hill, New York, 2002.

# 3

# Conditional Operators, Built-in Functions with Vector Arguments, MATLAB®'s Interp1 Function, and Some Scalar and Vector Operations

## 3.1 Introduction

In this chapter, we cover the next building block in basic programming, one that exists in most programming languages and that is the Conditional Operators. The first conditional operator discussed is the `if-else` command. Next we cover the `if-elseif-else` ladder, followed by a description of the `Switch Group`. Also, in this chapter, we discuss working with built-in functions with vector arguments, MATLAB®'s `interp1` function and some scalar and vector operations, including element-by-element operations.

## 3.2 Conditional Operators and Alternate Paths

### 3.2.1 The `if` Command Provides Two Alternate Paths

Syntax:

```
if logical expression
    statement;
       ⋮
    statement;
else
    statement;
       ⋮
    statement;
end
```

If the logical expression is true, then only the upper set of statements is executed. If the logical expression is false, then only the bottom set of statements is executed.

Logical expressions are of the form

```
a == b;     a <= b;
a < b;      a >= b;
a > b;      a ~= b;     (a not equal to b)
```

Logical expressions have only two values: true or false.

Compound logical expressions

```
a > b && a ~= c     (a > b and a ≠ c)
a > b || a < c      (a > b or a < c)
```

The following example illustrates the use of both the if command and the input command.

**NOTE 1:** In Example 3.1, we use the fprintf command just before the input command, to provide the user with more directions than can be given by the input command itself.

**NOTE 2:** When the input command is executed, the Run icon in the Editor Window changes to a Pause icon (see Figure 3.1).

**Example 3.1**

```
% Example_3_1.m
% This program uses the input command and an if statement to
% determine if the output is to go to the screen or to a file. The
% variables y1 and y2 are made vectors so that these variables can
% be printed outside the for loop that created them. As vectors,
% they can also be plotted.
clear; clc;
t=0:0.5:5;
```

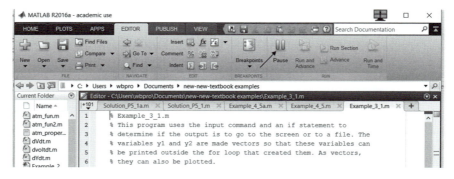

**FIGURE 3.1**
MATLAB's menu push button figure.

## Conditional Operators, Built-in Functions with Vector Arguments

```
for j=1:length(t)
    y1(j)=t(j)^2/10;
    y2(j)=t(j)^3/100;
end
fprintf('Do you wish to print the output to \n');
fprintf('the screen or to a file? \n');
response=input('Enter s for screen or F for file \n','s');
% Note, since we entered 's' in the input statement, you do not
% need to enclose your answer in single quotation marks.
if response=='s'
% Table headings:
    fprintf(' t            y1           y2 \n');
    fprintf('--------------------------------\n');
    for j=1:length(t)
        fprintf('%3.1f      %10.3f     %10.3f \n', ...
        t(j),y1(j),y2(j));
    end
end
if response=='F'
    fo=fopen('output.txt','w');
    % Table headings:
    fprintf(fo,' t           y1           y2 \n');
    fprintf(fo,'--------------------------------\n');
    for j=1:length(t)
        fprintf(fo,'%3.1f      %10.3f     %10.3f \n',...
        t(j),y1(j),y2(j));
    end
end
```

---

**Program Results (either from the screen or from the file "output.txt"):**

```
 t      y1       y2
-------------------------
0.0    0.000    0.000
0.5    0.025    0.001
1.0    0.100    0.010
1.5    0.225    0.034
2.0    0.400    0.080
2.5    0.625    0.156
3.0    0.900    0.270
3.5    1.225    0.429
4.0    1.600    0.640
4.5    2.025    0.911
5.0    2.500    1.250
-------------------------
```

### 3.2.2 The `if-elseif-else` Command Provides More than Two Alternate Paths

Syntax:

```
if logical expression 1
    statement(s);
elseif logical expression 2
    statement(s);
```

```
elseif logical expression 3
    statement(s);
else
    statement(s);
end
```

The `if-elseif-else` ladder works from top down. If the top logical expression is true, the statements related to that logical expression are executed, and the program will leave the ladder. If the top logical expression is not true, the program moves to the next logical expression. If that logical expression is true, the program will execute the group of statements associated with that logical expression and leave the ladder. If that logical expression is not true, the program moves to the next logical expression and continues the process. If none of the logical expressions are true the program will execute the statements associated with the `else` statement. The `else` statement is not required. In that case, if none of the logical expressions are true, no statements within the ladder will be executed.

---

Suppose in Example 3.1, we had more than two choices. For example, we had the choice of printing to the screen, printing to a file, exiting the program, or opening the edit window to create a new program. The following program uses the `if-elseif-else` ladder and the input statement to decide which choice to select.

**Example 3.2**

```
% Example_3_2.m
% First the script determines y1 and y2 as functions of t. The
% script then uses the input command and the if-elseif ladder to
% determine if the program prints the results to the screen,
% prints the results to a file, exits the program or opens the edit
% window to create a new script.
clear; clc;
t=0:0.5:5;
for j=1:length(t)
    y1(j)=t(j)^2/10;
    y2(j)=t(j)^3/100;
end
fprintf('If you wish to print the output to the screen enter S \n');
fprintf('If you wish to print the output to a file enter F \n');
fprintf('if you wish to close the program enter C \n');
fprintf('If you wish to open the edit window enter E \n');
fprintf('Enter your response without single quotation marks \n');
response=input('Enter S, F, C or E \n','s');
if response=='S'
% Table headings:
    fprintf('    t        y1        y2   \n');
    fprintf('   ------------------------------\n');
```

## Conditional Operators, Built-in Functions with Vector Arguments 85

```
    for j=1:length(t)
        fprintf('%5.1f     %10.3f     %10.3f \n',t(j),y1(j),y2(j));
    end
elseif response=='F'
    fo=fopen('output.txt','w');
    % Table headings:
    fprintf(fo,' t         y1        y2  \n');
    fprintf(fo,'-----------------------------\n');
    for j=1:length(t)
        fprintf(fo,'%5.1f     %10.3f     %10.3f \n',t(j),y1(j),y2(j));
    end
    fclose(fo);
elseif response=='C'
    quit;
elseif response=='E'
    edit;
end
```
---

### 3.2.3 The `break` Command

The `break` command may be used with an `if` statement to end a loop; Example:

```
for m = 1:20
    statement(s);
    if m > 10
        break;
    end
end
```

In the above example, when m becomes greater than 10, the program leaves the `for` loop and moves on to the next statement outside the for loop.

---

Frequently when we deal with problems involving material properties, we will find that there exist tables that describe several material properties of several different substances. For example, in Thermodynamics, there are textbooks that contain tables of the thermodynamic properties of specific volume, internal energy, and entropy of saturated water, ammonia, refrigerant 22, and so on as a function of temperature. Suppose we have a problem involving a material property that lies between table values. The simplest way to determine the material property would be to interpolate between table values. If we assume that the properties vary linearly between table values, then we can use linear interpolation. Suppose we have a table of $y$ as a function of $x$ and we wish to determine the value of $y$ at $x$, where $x$ lies between table values $x_1$ and $x_2$, then the general linear interpolation formula, based on similar triangles, in terms of $y$ and $x$ is

$$y = y_1 + \frac{(y_2 - y_1) \times (x - x_1)}{x_2 - x_1} \tag{3.1}$$

where $y_1$ and $y_2$ are the values of $y$ at $x_1$ and $x_2$, respectively.

Let us consider the Atmospheric problem described in Example 2.7. In that example, the atmospheric properties of temperature, pressure, and density as a function of altitude were specified in the file named *atm_properties.txt*, which is shown below. Here, the first column is altitude (m), the second column is temperature (K), the third column is pressure (Pa), and the fourth column is density (kg/m³).

| 0    | 288.15 | 1.0133e+005 | 1.2252 |
| 1000 | 281.65 | 8.9869e+004 | 1.1118 |
| 2000 | 275.15 | 7.9485e+004 | 1.0065 |
| 3000 | 268.65 | 7.0095e+004 | 0.9091 |
| 4000 | 262.15 | 6.1624e+004 | 0.8191 |
| 5000 | 255.65 | 5.4002e+004 | 0.7360 |

The next example is an interactive program, where the user is asked to enter an altitude at which he/she wishes to know the atmospheric properties. The program uses the `if-elseif` ladder to determine the closest surrounding altitude to the entered altitude. It then calculates the atmospheric properties of temperature, pressure, and density by linear interpolation. In MATLAB, the easiest and most efficient way to solve this interpolation problem is to use MATLAB's interp1 function. However, if you do not have MATLAB available, but have availability a different computer programming platform used by engineers, then you would probably need to solve the interpolation problem by one of the two examples listed below. The second one, Example 3.4 is more efficient than Example 3.3 and should always be used in preference to Example 3.3. We give Example 3.3 as a demonstration of the use of the `if-elseif` ladder. Later we will use the MATLAB's `interp1` function to solve the problem.

**Example 3.3**

```
% Example_3_3.m
% This program loads data from a file named atm_properties.txt
% The program asks the user to enter an elevation at which atmospheric
% properties are to be determined by linear interpolation.
% The altitude range is from 0 to 5000 m.
```

*Conditional Operators, Built-in Functions with Vector Arguments*         87

```
% Then the atmospheric properties are printed to the screen.
% The program uses the if-elseif ladder to select the closest interval
% to the entered altitude. The properties in this interval will be used
% in the interpolation formula.
% Temperature is in degrees Kelvin (K), pressure is in Pascal (Pa) and
% density is in kg/m^3.
clear; clc;
load atm_properties.txt
% establishing variable names to loaded data.
zt=atm_properties(:,1);
Tt=atm_properties(:,2);
pt=atm_properties(:,3);
rhot=atm_properties(:,4);
fprintf('Enter the altitude at which atmospheric properties \n');
z=input('are to be determined. Altitude range is from 0 to 5000 m \n');
if z>=zt(1)&& z<zt(2)
    z1=zt(1); z2=zt(2); T1=Tt(1); T2=Tt(2); p1=pt(1); p2=pt(2);
    rho1=rhot(1); rho2=rhot(2);
elseif z>=zt(2)&& z<zt(3)
    z1=zt(2); z2=zt(3); T1=Tt(2); T2=Tt(3); p1=pt(2); p2=pt(3);
    rho1=rhot(2); rho2=rhot(3);
elseif z>=zt(3)&& z<zt(4)
    z1=zt(3); z2=zt(4); T1=Tt(3); T2=Tt(4); p1=pt(3); p2=pt(4);
    rho1=rhot(3); rho2=rhot(4);
elseif z>=zt(4)&& z<zt(5)
    z1=zt(4); z2=zt(5); T1=Tt(4); T2=Tt(5); p1=pt(4); p2=pt(5);
    rho1=rhot(4); rho2=rhot(5);
elseif z>=zt(5)&& z<zt(6)
    z1=zt(5); z2=zt(6); T1=Tt(5); T2=Tt(6); p1=pt(5); p2=pt(6);
    rho1=rhot(5); rho2=rhot(6);
end
T=T1+(T2-T1)*(z-z1)/(z2-z1);
p=p1+(p2-p1)*(z-z1)/(z2-z1);
rho=rho1+(rho2-rho1)*(z-z1)/(z2-z1);
fprintf('T=%6.2f(K), p=%10.4e(Pa) rho=%6.4f(kg/m^3) \n',T,p,rho);
```
-----------------------------------------------------------------

**Program Results:**

```
Enter the altitude at which atmospheric properties
are to be determined. Altitude range is from 0 to 5000 m
4380
T=259.68(K), p=5.8728e+04(Pa) rho=0.7875(kg/m^3)
>>
```
-----------------------------------------------------------------

Whenever one gets the results of a program, it is prudent to examine the results to see if they make sense. In this case, do the obtained properties lie within the proper interval?

An alternative to loading the data in the file *atm_properties.txt* into the above script is to enter the data directly into the program as vectors. To accomplish this, replace the following lines in Example 3.3

```
load atm_properties.txt
% establishing variable names to loaded data.
zt=atm_properties(:,1);
Tt=atm_properties(:,2);
pt=atm_properties(:,3);
rhot=atm_properties(:,4);
```

with

```
zt=[0 1000 2000 3000 4000 5000];
Tt=[288.15 281.65 275.15 268.65 262.15 255.65];
pt=[10.133 8.9869 7.9485 7.0095 6.1624 5.4002]*1.0e+004;
rhot=[1.2252 1.1118 1.0065 0.9091 0.8191 0.7360];
```

A more efficient way to solve the problem with far fewer lines of code is to use a single for loop and an if statement to determine the closest interval to the entered altitude by the user, thus, reducing the number of lines in the program. This is demonstrated in the following example. This becomes important when the number of conditions in the program is large.

**Example 3.4**

```
% Example_3_4.m
% This program enters the data shown in atm_properties.txt directly
% into the program as vectors.
% The program then asks the user to enter an elevation at which the
% atmospheric properties are to be determined by linear interpolation.
% The atmospheric properties are then printed to the screen.
% The program uses a for loop and a compound if statement to determine
% the closest interval to the entered altitude. The properties in
% this interval will be used in the interpolation formula.
% Temperature is in degrees Kelvin (K), pressure is in Pascal (Pa) and
% density is in kg/m^3.
clear; clc;
zt=[0 1000 2000 3000 4000 5000];
Tt=[288.15 281.65 275.15 268.65 262.15 255.65];
pt=[10.133 8.9869 7.9485 7.0095 6.1624 5.4002]*1.0e+004;
rhot=[1.2252 1.1118 1.0065 0.9091 0.8191 0.7360];
fprintf('Enter the altitude at which atmospheric properties \n');
z=input('are to be determined. Altitude range is from 0 to 5000 m \n');
for i=1:length(zt)-1
    if z>=zt(i) && z<zt(i+1)
        z1=zt(i); z2=zt(i+1); T1=Tt(i); T2=Tt(i+1);
        p1=pt(i); p2=pt(i+1); rho1=rhot(i); rho2=rhot(i+1);
        break;
    end
end
T=T1+(T2-T1)*(z-z1)/(z2-z1);
p=p1+(p2-p1)*(z-z1)/(z2-z1);
rho=rho1+(rho2-rho1)*(z-z1)/(z2-z1);
fprintf('T=%6.2f(K), p=%10.4e(Pa)  rho=%6.4f(kg/m^3) \n',T,p,rho);
```

## Program Results:

```
Enter the altitude at which atmospheric properties
are to be determined. Altitude range is from 0 to 5000 m
1350
T=279.38(K), p=8.6235e+04(Pa) rho=1.0749(kg/m^3)
>>
```

---

### 3.2.4 The `switch` Command

In some cases, the Switch Group may be used as an alternative to the if-elseif-else ladder.

Syntax:

```
switch(var)
    case var1
        statement(s);
    case var2
        statement(s);
    case var3
        statement(s);
    otherwise
        statement(s);
end
```

where `var` takes on the possible values var1, var2, var3, and so on.

If var equals var1, those statements associated with var1 are executed and the program leaves the Switch Group. If var does not equal var1, the program tests if var equals var2, and if yes, the program executes those statements associated with var2 and leaves the Switch Group. If var does not equal any of var1, var2, and so on, the program executes the statements associated with the `otherwise` statement. If var1, var2, and so on are strings, they need to be enclosed by single quotation marks. It should be noted that var cannot be a logical expression, such as var1 > = 80.

The following example illustrates the use of the Switch Group in a MATLAB program.

### Example 3.5

```
% Example_3_5.m
% This program is a test of the switch statement.
clear; clc;
var = 'a';
x = 5;
switch(var)
    case 'b'
        z = x^2;
    case 'a'
        z = x^3;
```

```
        otherwise
            z=0;
end
fprintf(' z = %6.1f \n',z);
```
--------------------------------------------------------------------

**Program Results:**
```
z = 125.0
>>
```
--------------------------------------------------------------------

### 3.2.5 MATLAB's menu Function

MATLAB's menu function allows the user to select from several choices by the use of push buttons on a graphical display. Each item listed in the menu is given a number according to its position in the menu list. The top item in the menu display is given number 1, the second from the top is given number 2, and so on. In the following example, the user is prompted to click on one of the gases listed in the menu display (see Figure 3.2). The program then determines the density of the gas selected based on the ideal gas law. Pressure and temperature are specified.

**Example 3.6**
```
% Example_3_6.m
% This program uses MATLAB's menu function and the if-elseif ladder
% to determine the density of a gas by the ideal gas law.
% The gas is selected by the user by clicking on a push button in
% the menu display.
clear; clc;
p=2*1.013e+5; T=350.0;
Rt=[287 2077 4121 297 260];
k= menu('choose a gas','air','helium','hydrogen','nitrogen','oxygen');
if k==1
    R=Rt(1);
    fprintf('The gas is Air \n');
elseif k==2
    R=Rt(2);
    fprintf('The gas is Helium \n');
elseif k==3
    R=Rt(3);
    fprintf('The gas is hydrogen \n');
elseif k==4
    R=Rt(4);
    fprintf('The gas is Nitrogen \n');
elseif k==5
    R=Rt(5);
    fprintf('The gas is Oxygen \n');
end
rho=p/(R*T);
fprintf('The density, rho, is based on the ideal gas law \n');
fprintf('T=%5.1fK p=%10.4ePa rho=%7.4fkg/m^3 \n',T,p,rho);
```
--------------------------------------------------------------------

# Conditional Operators, Built-in Functions with Vector Arguments

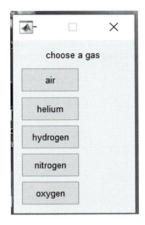

**FIGURE 3.2**
Menu display.

The menu display that pops up (upper left corner of the screen) is shown in Figure 3.2. I clicked on the nitrogen button and got the following result:

**Program Results:**
```
The gas is Nitrogen
The density, rho, is based on the ideal gas law
T=350.0K p=2.0260e+05Pa rho=1.9490kg/m^3
>>
```

---

## REVIEW 3.1

1. What statement is frequently used to establish two conditional paths?
2. What series of statements is used to establish several conditional paths?
3. List the various types of logic statements that can be used with the `if-else` and `if-elseif-else` ladder.
4. Is the `else` statement required with either the `if-else` or the `if-elseif-else` ladder?
5. What statement group and a MATLAB's function are alternatives to the `if-elseif-else` ladder?

**TABLE 3.1**

Score Range for Letter Grade

| Letter Grade | Score Range |
| --- | --- |
| A | 100 to 90 |
| B | 89 to 80 |
| C | 79 to 70 |
| D | 69 to 60 |
| F | Less than 60 |

**Exercises**

**E3.1.** The numerical grades on an exam are listed in the vector labeled scores.
Scores = [92 85 73 83 75 85 65 76 96 84 76 81 55 76 94 65 84 62 78 80 74 62 84 76 70 88 74 82 70 86]. Each score will be assigned a letter grade as indicated in the Table 3.1.

Write a MATLAB program that will determine the number of grades in each letter grade category and plot the result as a bar chart (see Example 2.14).

**E3.2.** Repeat Example 3.6, but this time use the `switch` Statement instead of the `if-elseif` ladder.

## 3.3 Working with Built-in Functions with Vector Arguments

MATLAB allows the built-in functions such as sin( ), cos( ), exp( ), and so on as well as functions in general to have vectors as arguments. **The result will also be a vector**. This is demonstrated in the next example.

**Example 3.7**

```
% Example_3_7.m
% This program demonstrates that if the argument in a built in
% function, such as MATLAB's sine function, is a vector, the result
% will also be a vector.
clear; clc;
% Define vector x;
x=0:30:360;
% Let y1 be the sine of a vector x where x is in degrees.
% Running sind with vector x as an argument will return a vector:
y1 = sind(x);
% Thus, y1 is a vector.
% Let y2(n) be the sine of the nth element of x. We will use a for
% loop to calculate each value y2(n) and then compare y1 and y2.
```

```
for n=1:length(x)
    y2(n)=sind(x(n));
end
% Table headings
fprintf('      x              y1              y2   \n');
fprintf('------------------------------------------\n');
for n=1:length(x)
    fprintf('%5.1f        %8.5f        %8.5f \n',x(n),y1(n),y2(n))
end
```

**Program Results:**

```
    x              y1              y2
------------------------------------------
   0.0         0.00000         0.00000
  30.0         0.50000         0.50000
  60.0         0.86603         0.86603
  90.0         1.00000         1.00000
 120.0         0.86603         0.86603
 150.0         0.50000         0.50000
 180.0         0.00000         0.00000
    .              .               .
    .              .               .
```

In the generated output, does y1 = y2?

We see that in some scripts, we could replace the use of a `for` loop by using a vector argument in many built-in or self-written functions, which produces a vector result, thus reducing the number of lines in the script. This concept was demonstrated in the above Example 3.7.

## 3.4 MATLAB's `interp1` Function

MATLAB has a function named interp1 that performs interpolation.
The syntax for `interp1` is

```
Yi = interp1(X,Y,Xi)
```

where X and Y are a set of known (x, y) data points and Xi is the set of x values at which the set of y values, Yi, are to be determined by linear interpolation. Arrays X and Y must be of the same length. Note: If Xi is a vector, then Yi will also be a vector. The function interp1 can also be used for interpolation methods other than linear interpolation, and this is covered in Chapter 9 on Curve Fitting. In the next example, we modify Example 3.4 by using MATLAB's interp1 function to interpolate for the atmospheric properties at an altitude entered from the keyboard.

## Example 3.8

```
% Example_3_8.m
% This program enters the data shown in atm_properties.txt directly
% into the program as vectors.
% The program then asks the user to enter an elevation at which the
% atmospheric properties are to be determined by linear interpolation.
% The atmospheric properties are then printed to the screen.
% The program uses MATLAB's interp1 function to do the interpolation.
% Temperature is in (K), pressure is in (Pa) and
% density is in (kg/m^3).
clear; clc;
zt=[0 1000 2000 3000 4000 5000];
Tt=[288.15 281.65 275.15 268.65 262.15 255.65];
pt=[10.133 8.9869 7.9485 7.0095 6.1624 5.4002]*1.0e+004;
rhot=[1.2252 1.1118 1.0065 0.9091 0.8191 0.7360];
fprintf('Enter the altitude at which atmospheric properties \n');
z=input('are to be determined. Altitude range is from 0 to 5000 m \n');
T=interp1(zt,Tt,z);
p=interp1(zt,pt,z);
rho=interp1(zt,rhot,z);
fprintf('T=%6.2f(K), p=%10.4e(Pa) rho=%6.4f(kg/m^3) \n',T,p,rho);
```
-----

**Program Results:**
```
Enter the altitude at which atmospheric properties
are to be determined. Altitude range is from 0 to 5000 m.
1350
T=279.38(K), p=8.6235e+04(Pa) rho=1.0749(kg/m^3)
>>
```

Are the results the same as those obtained in Example 3.4?
-----

The next example demonstrates the use of `interp1` function for interpolating for internal energy of a refrigerant at temperatures specified in vector T2. The output from interp1 will also be a vector.

## Example 3.9

```
% Example_3_9.m
% This program uses MATLAB's function interp1 to interpolate for
% the internal energy, u, as a function of temperature, T, of
% an unspecified refrigerant.
% Measured values of u in (kJ/kg) vs. T in degrees (C)
% are specified in vectors ut and Tt respectively.
% The temperatures at which the internal energy is to be determined
% are specified in vector T2.
% The program also creates a plot of u vs. T and includes points of
% u at temperature T2.
clear; clc;
Tt=-20:10:90;
ut=[217.86 224.97 232.24 239.69 247.32 255.12 263.10 271.25 ...
    279.58 288.08 296.75 305.58];
fprintf('This program interpolates for the internal energy, u at \n');
fprintf('a specified temperature T. \n');
```

# Conditional Operators, Built-in Functions with Vector Arguments

```
fprintf('The allowable temperature range is -20 to +90 C. \n\n');
T2=[-12 6 24 32 64 82];
u=interp1(Tt,ut,T2);
fprintf('    T2(C)      u(kJ/kg)   \n');
fprintf('-------------------------\n');
for i=1:length(T2)
    fprintf('  %6.1f       %8.3f \n',T2(i),u(i));
end
plot(Tt,ut,T2,u,'o' );
xlabel('T(C)'), ylabel('u(kJ/kg)'), title('u vs. T'), grid;
```

**Program Results:**

```
This program interpolates for the internal energy, u at
a specified temperature T.
The allowable temperature range is -20 to +90 C.

   T2(C)        u(kJ/kg)
--------------------------
  -12.0         223.548
    6.0         236.710
   24.0         250.440
   32.0         256.716
   64.0         282.980
   82.0         298.516
>>
```

See Figure 3.3

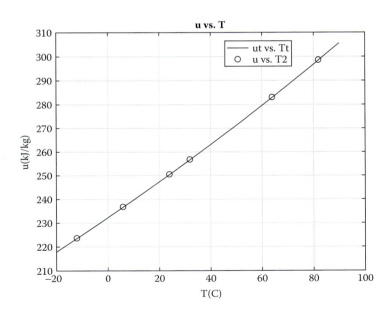

**FIGURE 3.3**
Internal energy, u, as a function of temperature, T, of an unspecified refrigerant.

## 3.5 Some Scalar and Vector Operations

### 3.5.1 Addition of a Scalar and a Vector

The result of a scalar added to a vector is that the scalar is added to each element of the vector. Example:

$$Z = 5 + [2\ 4\ 6\ 8] = [7\ 9\ 11\ 13]$$

### 3.5.2 Multiplication of a Scalar Times a Vector

The result of a scalar multiplied by a vector is that each element of the vector is multiplied by the scalar. Example:

$$Z = 5 * [2\ 4\ 6\ 8] = [10\ 20\ 30\ 40]$$

### 3.5.3 Addition and Subtraction of Two Vectors of the Same Length

Given $\mathbf{A} = [a_1\ \ a_2\ \ a_3]$ and $\mathbf{B} = [b_1\ \ b_2\ \ b_3]$, then

$$C = A + B \text{ gives } C = [(a_1 + b_1)\ (a_2 + b_2)\ (a_3 + b_3)]$$
$$D = A - B \text{ gives } D = [(a_1 - b_1)\ (a_2 - b_2)\ (a_3 - b_3)]$$

The addition or subtraction of two vectors of different lengths is not defined.

### 3.5.4 Element-by-Element Operations

Given two vectors of the same length, we can perform element-by-element multiplication, division, and exponentiation in MATLAB with the .*, and ./, and .^ operators.

Given $\mathbf{A} = [a_1\ \ a_2\ \ a_3]$ and $\mathbf{B} = [b_1\ \ b_2\ \ b_3]$, then

$$C = \mathbf{A}\ .*\ \mathbf{B} = [a_1 b_1\ \ a_2 b_2\ \ a_3 b_3],$$

$$D = \mathbf{A}\ ./\ \mathbf{B} = \left[ \frac{a_1}{b_1}\ \ \frac{a_2}{b_2}\ \ \frac{a_3}{b_3} \right],$$

$$E = \mathbf{A}\ .\wedge\ \mathbf{B} = [a_1^{b_1}\ a_2^{b_2}\ a_3^{b_3}] = power(A,B)$$

We see that the element-by-element operation results in a vector that is the same length as the vectors that are involved in the operation.

## Conditional Operators, Built-in Functions with Vector Arguments

**Example 3.10**

To demonstrate the above relations, copy the following script and run the program and observe the result.

```
% Example_3_10.m
% This program demonstrates some scalar and vector operations.
clear; clc;
s=5; a=[1 5 9]; b=[2 6 12]; c=s+a; d=s*b; e=a+b; f=a-b; g=a.*b;
h=a./b;
fprintf('s=%3i \n',s);
fprintf('a= %3i %3i %3i \n',a);
fprintf('\n');
fprintf('c=s+a\n');
fprintf('c= %3i %3i %3i \n',c);
fprintf('\n');
fprintf('s=%4i \n',s);
fprintf('b= %3i %3i %3i \n',b);
fprintf('d=s*b \n');
fprintf('d= %3i %3i %3i \n',d);
fprintf('\n');
fprintf('a= %3i %3i %3i \n',a);
fprintf('b= %3i %3i %3i \n',b);
fprintf('e=a+b \n');
fprintf('e= %3i %3i %3i \n',e);
fprintf('\n');
fprintf('a= %3i %3i %3i \n',a);
fprintf('b= %3i %3i %3i \n',b);
fprintf('f=a-b \n');
fprintf('f= %3i %3i %3i \n',f);
fprintf('\n');
fprintf('a= %3i %3i %3i \n',a);
fprintf('b= %3i %3i %3i \n',b);
fprintf('g=a.*b \n');
fprintf('g= %3i %3i %3i \n',g);
fprintf('\n');
fprintf('a= %3i %3i %3i \n',a);
fprintf('b= %4i %4i %4i \n',b);
fprintf('h=a./b \n');
fprintf('h= %8.4f %8.4f %8.4f \n',h);
fprintf('\n');
a=[2 3 4]; b=[2 2 2];
k=a.^b;
fprintf('a= %3i %3i %3i \n',a);
fprintf('b= %4i %4i %4i \n',b);
fprintf('k=a.^b \n');
fprintf('k= %4.1f %4.1f %4.1f \n',k(1),k(2),k(3));
```
------------------------------------------------------------------

**Program Results:**
```
s=   5
a=   1   5   9
c=s+a
c=   6  10  14
```

```
s=    5
b=    2    6    12
d=s*b
d=    10   30   60

a=    1    5    9
b=    2    6    12
e=a+b
e=    3    11   21

a=    1    5    9
b=    2    6    12
f=a-b
f=   -1   -1   -3

a=    1    5    9
b=    2    6    12
g=a.*b
g=    2    30   108

a=    1    5    9
b=    2    6    12
h=a./b
h=    0.5000    0.8333    0.7500

A=    2    3    4
B=    2    2    2
k=A.^B
k=    4.0   9.0   16.0
>>
```

---

### 3.5.5 Operation of Two Vector Functions

If a script involves a mathematical operation of two vector functions (such as a product of two vector functions), then the operation will require an element-by-element operation. In Example 3.9, we compute the product of two vector functions, both directly and indirectly by using a `for` loop and multiplying the elements of each vector. We then compare the results.

**Example 3.11**

```
% Example_3_11.m
% This example illustrates element-by element operation
% of two vector functions
clear; clc;
x = 0:30:180;
% y1 is the product of two vector functions
y1 = sind(x).* cosd(x);
fprintf('    x           y1          y2   \n');
fprintf('---------------------------------------\n');
for n=1:length(x)
    % y2(n) is the product of the elements of the two functions.
```

## Conditional Operators, Built-in Functions with Vector Arguments

```
      y2(n) = sind(x(n)) * cosd(x(n));
      fprintf('%5.1f     %8.5f     %8.5f \n',x(n),y1(n),y2(n));
end
```

---

**Program Results:**

```
     x           y1              y2
-----------------------------------------
   0.0        0.00000         0.00000
  30.0        0.43301         0.43301
  60.0        0.43301         0.43301
  90.0        0.00000         0.00000
 120.0       -0.43301        -0.43301
 150.0       -0.43301        -0.43301
 180.0       -0.00000        -0.00000
>>
```

---

We see that the two different methods for computing y1 and y2 give the same answer.

### REVIEW 3.2

1. If y = 3.0 * **A** and **A** is a vector, what can you say about y?
2. If y = 3.0 * sin(x) and x is a vector, what can you say about y?
3. If vector **C** = **A** + **B**, what must be true about vectors **A** and **B**.
4. What is the result of the multiplication of two vectors of the same length, say **A** and **B**, and how must it be programmed?
5. What is the name of MATLAB's function that does interpolation?
6. What are the inputs to MATLAB's interpolation function?
7. What are the outputs from MATLAB's interpolation function?

### Projects

**P3.1.** Though atmospheric conditions vary from day-to-day, it is convenient for design purposes, to have a model for atmospheric properties with altitude. The U.S. Standard Atmosphere, modified in 1976, is such a model. The model consists of two types of regions, one in which the temperature varies linearly with altitude, and the other is a region where the temperature is a constant (see Figure P3.1).

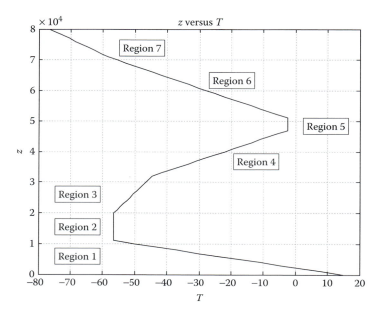

**FIGURE P3.1**
Temperature versus altitude of U.S. Standard Model Atmosphere.

The temperature and approximate pressure relations are as follows:

1. For a region where the temperature varies linearly

$$p = p_i \left( 1 - \frac{\lambda_i (z_{i+1} - z_i)}{T_i} \right)^{\frac{g_i}{\lambda_i R}} \qquad (P3.1)$$

$$T = T_i - \lambda_i (z_{i+1} - z_i) \qquad (P3.2)$$

where:
$z_i$ is the altitude at the beginning of region $i$, $i = 1, 2, \ldots, 7$
$z_{i+1}$ is the altitude at the end of region $i$ and at the beginning of region $i+1$
$(p_i, T_i)$ is the pressure and temperature at the beginning of region $i$
$\lambda_i$ is the lapse rate in region $i$
$R$ is the air gas constant = $286.9$ N–m/kg–K
$g_i$ is the gravitational constant in region $i$. Although $g$ varies slightly with altitude we take $g$ to be a constant in the region evaluated at $z_i$, otherwise the above expression for $p$ would be a lot more complicated than the one shown above

### Conditional Operators, Built-in Functions with Vector Arguments

2. For a region where the temperature is constant ($\lambda = 0$)

$$T = T_i \qquad \text{(P3.3)}$$

$$p = p_i \exp\left(-\frac{g_i(z-z_i)}{RT_i}\right) \qquad \text{(P3.4)}$$

For each region, the governing equation for $g$ is

$$g = g_o\left(1 - \frac{2z}{R_o}\right) \qquad \text{(P3.5)}$$

where:
$g_o = 9.810$ m/s²
$R_o = 6.3714\text{e}+6$ m
$z$ is the altitude at the beginning of the region

Create a MATLAB program that will

1. Determine the values of $T_i$, $p_i$, and $g_i$, $i = 1, 2, \ldots, 8$
2. Construct a table filling in the unknown values listed in Table P3.1.

**TABLE P3.1**
U.S. Standard Atmosphere Property Table

| Region | $z_i$ (km) | $T_i$ (K) | $p_i$ (Pa) | $\lambda_i$ (K/m) | $g_i$ (m/s²) |
|---|---|---|---|---|---|
|   | 0 | 288.15 | 101325 | — | 9.810 |
| 1 | — | — | — | 0.0065 | — |
|   | 11.0 | — | — | — | — |
| 2 | — | — | — | 0.0000 | — |
|   | 20.0 | — | — | — | — |
| 3 | — | — | — | −0.001 | — |
|   | 32.0 | — | — | — | — |
| 4 | — | — | — | −0.0028 | — |
|   | 47.0 | — | — | — | — |
| 5 | — | — | — | 0.0000 | — |
|   | 51.0 | — | — | — | — |
| 6 | — | — | — | 0.0028 | — |
|   | 71.0 | — | — | — | — |
| 7 | — | — | — | 0.0020 | — |
|   | 84.9 | — | — | — | — |

**P3.2.** This project is a modification of Example 3.8. Instead of making the program interactive, enter the following altitudes, z2, at which the atmospheric properties of $T, p, \rho$ are to be determined by linear interpolation using MATLAB's `interp1` function.

$$z2 = [680 \ 1250 \ 2360 \ 3685 \ 4320 \ 4865]$$

Print the results to a file in a table format.

**P3.3.** The properties of specific volume, v, and pressure, $p$, as a function of temperature, $T$, for carbon dioxide based on the Redlich–Kwong Equation of state are given in the tables below:

**NOTE:** 1 bar = $10^5$ N/m²

Write a MATLAB program that will do the following:

1. Construct three separate vectors containing the carbon dioxide variables of $T$, v, and $p$.
2. Print Table P3.2 to the screen (with table headings).
3. Determine v and $p$ at temperatures $T2$ using MATLAB's `interp1` function. Take

$$T2 = [367 \ 634 \ 420 \ 587 \ 742]$$

4. Print to the screen in a table format (with table headings) values of v and $p$ at temperatures $T2$.

**TABLE P3.2**

Properties of Carbon Dioxide Based on the Redlich–Kwong Equation of State

| T (K) | v (m³/kmol) | p (bar) |
|---|---|---|
| 350 | 0.28 | 7.65 |
| 400 | 0.32 | 8.57 |
| 450 | 0.36 | 9.16 |
| 500 | 0.40 | 9.55 |
| 550 | 0.44 | 9.81 |
| 600 | 0.48 | 10.00 |
| 650 | 0.52 | 10.14 |
| 700 | 0.56 | 10.24 |
| 750 | 0.60 | 10.31 |

### TABLE P3.3
### Refrigerant Properties

| T (°C) | v (m³/kg) | u (kJ/kg) |
|---|---|---|
| −35 | 0.34235 | 244.33 |
| −30 | 0.35369 | 217.58 |
| −20 | 0.36513 | 222.43 |
| −10 | 0.37649 | 227.37 |
| −5 | 0.35778 | 232.42 |
| 0 | 0.39901 | 237.57 |
| 10 | 0.41019 | 242.82 |
| 20 | 0.42133 | 248.16 |
| 25 | 0.43243 | 253.61 |
| 30 | 0.44348 | 259.16 |

**P3.4.** The measured properties of a refrigerant are shown in Table P3.3. Create a MATLAB program that will **repeatedly** ask the user if he/she wishes to determine the refrigerant properties. If the answer is Y for yes, then ask the user to enter a temperature at which the refrigerant properties are to be determined. Use MATLAB's interp1 function to evaluate the refrigerant properties, and then print the result to the screen.

**P3.5.** This project involves the mass-spring-dashpot system (see Figure 2.19). If the mass is displaced from its equilibrium position and released, the subsequent motion of the mass will depend on the values of $m$, $k$, and $c$. To simplify the writing of the governing equations, we will define

$$\text{arg} = \frac{k}{m} - \left(\frac{c}{2m}\right)^2$$

If arg > 0, then the displacement, $y$, as a function of time, $t$, will be

$$y = \exp\left(-\frac{ct}{2m}\right) \times \left(A\cos\left(\sqrt{\text{arg}}\ t\right) + B\sin\left(\sqrt{\text{arg}}\ t\right)\right)$$

If arg < 0, then the displacement, $y$, as a function of time, $t$, will be

$$y = \exp\left(-\frac{ct}{2m}\right) \times \left(A\exp\left(\sqrt{-\text{arg}}\ t\right) + B\exp\left(-\sqrt{-\text{arg}}\ t\right)\right)$$

where $\exp(x) = e^x$
   We wish to consider four cases.

   a. $m = 75$ kg, $k = 85$ N/m, $c = 200$ N-s/m.
   b. $m = 75$ kg, $k = 150$ N/m, $c = 40$ N-s/m.
   c. $m = 75$ kg, $k = 50$ N/m, $c = 150$ N-s/m.
   d. $m = 75$ kg, $k = 200$ N/m, $c = 20$ N-s/m.

Create a MATLAB program that will plot the motion of $y$ as a function of $t$ for $0 \le t \le 20$s in steps of 0.1 s. To make the initial displacement equal to 0.5 m and initial velocity equal to zero, use the following values for $A$ and $B$.
   For the case arg > 0, take

$$A = 0.5 \text{ m}, \quad B = \frac{Ac}{2m\sqrt{\text{arg}}}$$

For the case arg < 0, take

$$A = \frac{0.5\left(\sqrt{-\text{arg}} + \dfrac{c}{2m}\right)}{2\sqrt{-\text{arg}}}, \quad B = A\frac{\sqrt{-\text{arg}} - \dfrac{c}{2m}}{\sqrt{-\text{arg}} + \dfrac{c}{2m}}$$

# 4

# Self-Written Functions and MATLAB®'s `fminbnd` Function

## 4.1 Introduction

In this chapter, we cover the self-written function, which is the last of the building blocks that we mentioned earlier. Self-written functions are useful if you have a complicated program and wish to break it down into smaller segments. Also, if a series of statements is to be used many times, it is convenient to place them in a function. Many MATLAB® functions (such as `fminbnd`, `fzero`, `quad`, and `ode45`) require a self-written function to define the problem of interest. Self-written functions are equivalent to subroutines in most programming languages, but in MATLAB they are usually stored in separate files instead of the main program (though small functions can be defined in the same file as your main script, as described in the section on anonymous functions). The function file name must be saved as a .m file.

MATLAB has a function that determines the relative minimum or relative maximum of a single variable function. The function is `fminbnd` and is covered in this chapter. This MATLAB function requires the user to write a self-written function, either as a separate .m file or as an anonymous function within the main script.

## 4.2 Self-Written Function

MATLAB has a template for writing a function (see Figure 4.1), which can be accessed by clicking on the *down arrow under new* in the toolstrip in the Editor Window and selecting function from the dropdown menu. To create your function, you would need to edit the function template and make the desired changes. However, it is just as easy to open a new script window and type in your desired function and save it with the function name and as a .m file. The function template is of the form

```
function [output arguments] = function_name (input arguments)
```

*105*

**FIGURE 4.1**
MATLAB's function template.

**TABLE 4.1**

Example Function Definitions

| Function Definition Line | Function File Name |
|---|---|
| function [p,T] = atm(z,rho) | atm.m |
| function ex = exf (x) | efx.m |
| function[] = output(x,y) | output.m |

*The first executable statement in the function file must start with the word function.* Some example function definitions are shown in Table 4.1.

If the function has more than one output value, then the output variables must be in brackets. If there is only one output value, then no brackets are necessary. If there are no output values, use empty brackets.

1. *Input and Output Arguments*:

    *Input and output arguments in the function may be either scalars, vectors, or matrices. The input arguments defined in the calling program passes information to the function, where it is most often used in one or more arithmetic statements. The input arguments in the calling program need to be defined before the function is called. The output arguments determined in the function passes information to the calling program.*

## Self-Written Functions and MATLAB®'s fminbnd Function 107

2. *Variables Defined and Manipulated inside the Function*:

*Variables defined and manipulated inside the function are local to the function. This means that the only communication between the calling program and the function is through the input and output arguments of the function. The exception to this is when a* `global` *statement is contained in both the calling program and in the function.*

In the next example, the `input` command is again used to ask the user to enter an altitude from the keyboard. Earlier, it was mentioned that MATLAB version R2016a has a bug. When the `input` command is executed, the cursor remains in the Editor Window instead of moving to the Command Window. So the user needs to click in the Command Window before entering an altitude. This bug will be fixed in the next version of MATLAB.

### Example 4.1, Part A

```
% Example_4_1_parta.m
% The program asks the user to enter an elevation at which the
% atmospheric properties are to be determined by linear interpolation.
% The function atm_fun contains atmospheric properties every
% thousand meters and does the interpolation by MATLAB's
% interp1 function. The properties at the entered elevation is
% passed on to the calling program where it is printed to the
% Command window.
% In this example input and out variables are scalars.
% Temperature is in degrees Kelvin (K), pressure is in Pascal (Pa) and
% density is in kg/m^3.
clear; clc;
fprintf('Enter the altitude at which atmospheric properties are \n');
z=input('to be determined. Altitude range is from 0 to 5000 m \n');
[T,p,rho]=atm_fun(z);
fprintf('z=%6.1f(m)  T=%6.2f(K)  ',z,T);
fprintf('p=%10.4e(Pa)  rho=%6.4f(kg/m^3) \n',p,rho);
```
---

### Example 4.1, Part B

```
% atm_fun.m
% This function works with Example 4.1, part A
function [T,p,rho]=atm_fun(z)
zt=[0 1000 2000 3000 4000 5000];
Tt=[288.15 281.65 275.15 268.65 262.15 255.65];
pt=[10.133 8.9869 7.9485 7.0095 6.1624 5.4002]*1.0e+004;
rhot=[1.2252 1.1118 1.0065 0.9091 0.8191 0.7360];
T=interp1(zt,Tt,z);
p=interp1(zt,pt,z);
rho=interp1(zt,rhot,z);
```
---

**Program Results:**

```
Enter the altitude at which atmospheric properties
are to be determined. Altitude range is from 0 to 5000 m
3400
z=3400.0(m), T=266.05(K), p=6.6707e+04(Pa), rho=0.8731(kg/m^3)
>>
```

We now want to modify Example 4.1, so that the input and output variables to the function are vectors. We only need to modify Example 4.1, Part A. We do not need to modify the function atm_fun.

**Example 4.2, Part A**

```
% Example_4_2_parta.m
% This program specifies vector z as an input to function atm_fun.
% The function atm_fun contains atmospheric properties every
% thousand meters and does the interpolation by MATLAB's
% interp1 function at each element of vector z.
% The output variable of T, p and rho from atm_fun will be vectors.
% These vectors are passed on to the calling program where it is
% printed to the screen in table format.
% Temperature is in degrees Kelvin (K), pressure is in Pascal (Pa) and
% density is in kg/m^3.
clear; clc;
z=[1250 2560 3480 4360];
[T,p,rho]=atm_fun(z);
fprintf(' z(m)         T(K)         p(Pa)        rho(kg/m^3)   \n');
fprintf('-----------------------------------------------------\n');
for i=1:length(z)
    fprintf('%6.1f     %6.2f     %10.4e    %6.4f \n',...
            z(i),T(i),p(i),rho(i));
end
```

**Example 4.2, Part B is the same as Example 4.1, Part B**

**Program Results:**

```
 z(m)       T(K)        p(Pa)        rho(kg/m^3)
-----------------------------------------------
 1250.0     280.02      8.7273e+04     1.0855
 2560.0     271.51      7.4227e+04     0.9520
 3480.0     265.53      6.6029e+04     0.8659
 4360.0     259.81      5.8880e+04     0.7892
>>
```

The following example demonstrates that the names of the arguments in the calling program need not be the same as those in the function. It is only the order of the argument list in the calling program that needs to be in the

## Self-Written Functions and MATLAB®'s fminbnd Function

same order as the argument list defined in the function. This feature is useful when a function is to be used with several different scripts, each script having different variable names, but each of the variables names correspond to variables in the function. This concept is used in all of MATLAB's built-in functions.

Let us modify Example 4.2, Part A, and name it Example 4.3, Part A.

### Example 4.3, Part A

```
% Example_4_3_parta.m
clear; clc;
z=[1250 2560 3480 4360];
[T,p,rho]=atm_fun2(z);
fprintf('\n This output is from Example 4.3, Part A \n');
fprintf(' z(m)        T(K)         p(Pa)        rho(kg/m^3)   \n');
fprintf('-----------------------------------------------------\n');
for i=1:length(z)
    fprintf('%6.1f      %6.2f      %10.4e      %6.4f \n',...
            z(i),T(i),p(i),rho(i));
end
```
-----------------------------------------------------------------

Now we will modify function atm_fun and name it atm_fun2.

```
% atm_fun2.m
% This function works with Example 4.3, part A
function [A,B,C]=atm_fun2(x)
alt=[0 1000 2000 3000 4000 5000];
Temp=[288.15 281.65 275.15 268.65 262.15 255.65];
pres=[10.133 8.9869 7.9485 7.0095 6.1624 5.4002]*1.0e+004;
dens=[1.2252 1.1118 1.0065 0.9091 0.8191 0.7360];
A=interp1(alt,Temp,x);
B=interp1(alt,pres,x);
C=interp1(alt,dens,x);
fprintf('This output is from atm_fun2 \n');
fprintf(' x(m)        A(K)         B(Pa)        C(kg/m^3)   \n');
fprintf('-----------------------------------------------------\n');
for i=1:length(x)
    fprintf('%6.1f      %6.2f      %10.4e      %6.4f \n',...
            x(i),A(i),B(i),C(i));
end
```
-----------------------------------------------------------------

### Program Results:

```
This output is from atm_fun2
  x(m)        A(K)         B(Pa)        C(kg/m^3)
-----------------------------------------------------
  1250.0      280.02       8.7273e+04    1.0855
  2560.0      271.51       7.4227e+04    0.9520
  3480.0      265.53       6.6029e+04    0.8659
  4360.0      259.81       5.8880e+04    0.7892
```

```
This output is from Example 4.3, Part A
   z(m)          T(K)          p(Pa)        rho(kg/m^3)
------------------------------------------------------
  1250.0        280.02        8.7273e+04      1.0855
  2560.0        271.51        7.4227e+04      0.9520
  3480.0        265.53        6.6029e+04      0.8659
  4360.0        259.81        5.8880e+04      0.7892
>>
```

Comparing results, we see that the names of the arguments in the calling program need not be the same as those in the function. In the calling program, the names of the input and output arguments are z, T, p, and rho. In the function, atm_fun2, the names of the input and output arguments are x, A, B, and C. Looking at the results, we see that x = z, A = T, B = p, and C = rho. It is only the order of the argument list in the function that needs to be in the same order as the argument list in the calling program.

## 4.3 Anonymous Functions

Sometimes it is more convenient to define a function inside your script rather than in a separate file. For example, if a function is brief (perhaps a single line) and unlikely to be used in other scripts, then the *anonymous* form of a function can be used. This will save you from having to create another .m file. The syntax for an anonymous function is

```
funhandle = @(arg_list) (function expression)
```

A function handle is a MATLAB value that provides a means of calling a function indirectly. An example of an anonymous function is

```
fh = @(x,y) (y*sin(x)+x*cos(y));
```

In the above expression, MATLAB lists the @ sign as a function handle creation, fh is the function handle, the (x,y) defines the input arguments to the function, and (y*sin(x)+x*(cos(y)) is the function. Anonymous functions may be used in a script or in the Command Window.

Example: In the Command Window, type-in the following two lines:

```
>> fh = @(x,y) (y*sin(x)+x*cos(y));
>> w = fh(pi,2*pi)
     w =
           3.1416
```

Additional information on anonymous functions can be obtained by typing help function_handle in the Command Window.

# Self-Written Functions and MATLAB®'s fminbnd Function

**Example 4.4**

The following example uses an anonymous function that employs the interpolation formula described in Equation 3.1. The script is a modification of Example 3.4 that contains atmospheric data of temperature, pressure, and density every 1000 meters. The script, interactively, asks the user to enter an altitude from the keyboard.

```
% Example_4_4.m
% This program interpolates for atmospheric properties T, p and rho at
% an altitude entered from the keyboard.
% Atmospheric properties of temperature, pressure and density
% are specified every 1000 meters. The atmospheric properties at an
% altitude entered from the key board are determined by
% the interpolation formula described in Equation 3.1 and printed to
% the screen. An anonymous function, which avoids creating an extra .m
% file, is used to do the interpolation:
clear; clc;
% anonymous function
yf = @(z,z1,z2,y1,y2) (y1+(z-z1)*(y2-y1)/(z2-z1));
zt=[0 1000 2000 3000 4000 5000];
Tt=[288.15 281.65 275.15 268.65 262.15 255.65];
pt=[10.133 8.9869 7.9485 7.0095 6.1624 5.4002]*1.0e+004;
rhot=[1.2252 1.1118 1.0065 0.9091 0.8191 0.7360];
fprintf('Do you wish to have the atmospheric properties \n');
fprintf('at a specific altitude determined \n');
char=input('enter Y for yes or N for no \n','s');
if char=='N'
    fprintf('Good Bye \n');
    exit;
end
while char=='Y'
    fprintf('Enter the altitude at which atmospheric properties \n');
    fprintf('are to be determined. \n');
    z=input('Altitude range is from 0 to 5000 m \n');
    for i=1:length(zt)-1
        if z >= zt(i) && z < zt(i+1)
            z1=zt(i); z2=zt(i+1);
            T1=Tt(i); T2=Tt(i+1);
            T=yf(z,z1,z2,T1,T2);
            p1=pt(i); p2=pt(i+1);
            p=yf(z,z1,z2,p1,p2);
            rho1=rhot(i); rho2=rhot(i+1);
            rho=yf(z,z1,z2,rho1,rho2);
            fprintf('z=%6.1f(m)    T=%7.2f(C)',z,T);
            fprintf('p=%12.5e(Pa)  rho=%8.5f(kg/m^3) \n',p,rho);
        end
    end
    fprintf('\n');
    fprintf('Do you wish to enter another altitude ');
    char=input('enter Y for yes or N for no \n','s');
    if char=='N'
        fprintf('Good Bye \n');
    end
end
```

**Program Results:**

```
Do you wish to determine atmospheric properties
at a specific altitude,
enter Y for yes or N for no
Y
Enter the altitude at which atmospheric properties
are to be determined.
Altitude range is from 0 to 5000 m
1480
z=1480.0(m)  T= 278.53(C)  p= 8.48847e+04(Pa)  rho= 1.06126(kg/m^3)

Do you wish to enter another altitude enter Y for yes or N for no
Y
Enter the altitude at which atmospheric properties
are to be determined.
Altitude range is from 0 to 5000 m
3620
z=3620.0(m)  T= 264.62(C)  p= 6.48430e+04(Pa)  rho= 0.85330(kg/m^3)

Do you wish to enter another altitude enter Y for yes or N for no
N
Good Bye
>>
```

## REVIEW 4.1

1. When does it seem appropriate to write a self-written function?
2. In writing a self-written function what must be the first word in the first executable statement?
3. A self-written function usually has both an input and an output. Where does the input come from? Where does the output go to?
4. If a self-written function has more than one output, how must the output be presented?
5. How does a self-written function communicate with the calling program?
6. What can be said about variables in the self-written function that are not in the input or output arguments of the function and there are no `global` statements?
7. Do the variable names in the input and output arguments between the calling program and the function have to be the same?
8. If a programmer wishes to write a self-written function, but does not wish to create an additional .m file, what can the programmer do and what is the constraint?

## 4.4 MATLAB's fminbnd

There are times when we might be interested in determining a relative minimum of a single variable function. MATLAB has a built-in function that will do this for us. The syntax for the function is

```
[x,FVAL] = fminbnd(FUN,x1,x2)
```

The function fminbnd determines the relative minimum of a single variable function in the interval x1<x<x2. The FUN argument is a function handle to the function that describes the function whose relative minimum we wish to determine. The arguments x1, x2 give the interval in which the relative minimum may lie and the output x is the x value at the relative minimum. FVAL gives the functional value at x. FUN can be a function defined in a separate .m file or may be defined by an anonymous function or within the fminbnd function itself with the use of the function handle creator, @, as shown in Example 4.5. Note that MATLAB does not have a separate function to find a relative maximum. In order to find a relative maximum, redefine FUN to return the negative value of the function of interest, and then use fminbnd to find the relative minimum (see Example 4.5). This is also an example in which the user needs to write a self-written function in order to use a MATLAB built-in function.

**Example 4.5**

Given: $y(x) = x^3 + 5.7x^2 - 35.1x + 85.176$.
    Determine the relative minima and maxima.

```
% Example_4_5.m
% Find the minima and maxima of y = x^3 + 5.7x^2 - 35.1x + 85.176
clc; clear;
% First, plot the function so that we can determine the x range to use
% in fminbnd. Let us assume that the relative minimum lies between
% x1=-10, x2=6.
xf=-10:0.1:6;
y=xf.^3 + 5.7*xf.^2 - 135*xf + 85.176
plot(xf,y), xlabel('x'),ylabel('y'), grid, title('y vs x');
% Next, find the minimum and maximum using MATLAB's anonymous
% function method directly in the fminbnd function.
[xmin,ymin] = fminbnd( @(x) (x^3+5.7*x^2-35.1*x+85.176),-10,6 );
fprintf('xmin=%7.3f   ymin=%9.3f \n',xmin,ymin)
% Note: To find a maximum, instead find the minimum of the
% negative of the function.
```

```
[xmax,ymax] = fminbnd( @(x) -(x^3+5.7*x^2-35.1*x+85.176),-10,6);
Ymax=-ymax;
% Print results
fprintf('xmax=%7.3f Ymax=%9.3f \n',xmax,Ymax);
```
---

**Program Results:**

```
xmin= 2.013 ymin= 45.774
xmax= -5.813 Ymax= 285.394
>>
```

See Figure 4.2.
---

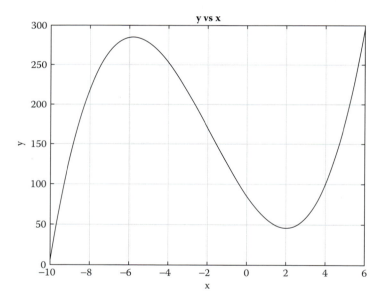

**FIGURE 4.2**
Relative maximum and minimum of y versus x.

# Self-Written Functions and MATLAB®'s `fminbnd` Function

## Projects

**P4.1.** Before scientific calculators and computers existed, numerical values for functions such as ln x, e$^x$, cos x, and so on were given in tables. The table values were determined by power series, such as a Maclaurin or a Taylor series. For example, the cosine function can be represented by the following series:

$$\cos x = 1 - \frac{x^2}{2!} + \frac{x^4}{4!} - \frac{x^6}{6!} + \dots \dots \frac{x^n}{n!} \dots \quad \text{valid for } [x^2 < \infty] \quad \text{(P4.1)}$$

where $n$ is an even number and $x$ is in radians. The project involves creating a MATLAB program that contains both a calling program and a function. The calling program is to *continuously* ask the user if he/she wishes to know the value of the cosine at a specific angle ranging from 0 to 360 degrees. If the answer is N for *no*, exit the program. If the answer is Y for *yes*, the calling program is to ask the user to enter an angle from the keyboard. The calling program is to convert the entered angle to radians, which is the x value to be used in Equation P4.1. The calling program then is to call the function using the x value as an input to the function. The output from the function should be the value of the cosine function evaluated at x by Equation P4.1. The calling program is then to print the angle, the x value, and the cosine value to the screen. The calling program is to repeat the process until the user responds with a N.

**P4.2.** Repeat Project 4.1, but this time use the series expansion for the sine function, which is

$$\sin x = \frac{x}{1!} - \frac{x^3}{3!} + \frac{x^5}{5!} - + \dots \dots \frac{x^n}{n!} \dots \quad \text{valid for } [x^2 < \infty] \quad \text{(P4.2)}$$

where $n$ is an odd number.

**P4.3.** In this project we consider the fluid level, $h$, in a tank, as it discharges through a small circular hole (orifice) of diameter, $d$, near the bottom of the tank (see Figure P2.4, page 75).

The tank has a circular cross section of diameter D. A formula describing the fluid level, $h$, in the tank is

$$h = h_o - \frac{C_d A_o}{A_T}\sqrt{2 g h_o}\; t + \left(\frac{C_d A_o}{A_T}\right)^2 \times \frac{g t^2}{2} \quad \text{(P4.3a)}$$

where:
$C_d$ is the discharge coefficient
$h_o$ is the fluid level in the tank at time, $t = 0$
$A_o$ is the area of the orifice
$A_T$ is the cross-sectional area of the tank

The discharge coefficient, $C_d$, for a particular tank and orifice is determined by experiment. Create a MATLAB program that consists of the following:

1. A calling program that calls a self-written function with *input* arguments of $A_o$, $A_T$, $h_o$, and $C_d$. Take $d = 0.0055$ m, $D = 0.146$ m, $h_o = 0.288$ m, $g = 9.81$ m/s² and $C_d = 0.6$, where $d$ and $D$ are the diameters of the orifice and tank, respectively.
2. A self-written function as a .m file with output argument vectors $h$ and $t$.

Take $t = 0$ to 200 seconds in steps of 4 seconds and $g = 9.81$ m/s².
In the calling program print to a file a table consisting of $h$ versus $t$, with table headings and units. Also plot a graph of $h$ versus $t$.

**P4.4.** This project is a variation of Project P4.3.

Create a MATLAB program that determines $h(t)$ by the use of an anonymous function for Equation P4.3a. Take $C_d = [0.5\ 0.6\ 0.7]$ and $t = 0$ to 200 seconds in steps of 4 seconds. For each $C_d$ create a table of $h$ versus $t$, with table headings. For each $C_d$ create a plot of $h$ versus $t$, all on the same page.

Use the same parameters that was used in Project P4.3, that is, take $d = 0.0055$ m, $D = 0.146$ m, $h_o = 0.288$ m, and $g = 9.81$ m/s².

**P4.5.** Several properties of a refrigerant as a function of temperature are shown below in Table P4.1:

**TABLE P4.1**

Refrigerant Properties

| T(°C) | v (m³/kg) | u(kJ/kg) |
|---|---|---|
| −20 | 0.31003 | 206.12 |
| −10 | 0.34992 | 224.97 |
| 0 | 0.36433 | 232.24 |
| 10 | 0.37861 | 239.69 |
| 20 | 0.39279 | 247.32 |
| 30 | 0.40688 | 255.12 |
| 40 | 0.42091 | 263.10 |
| 50 | 0.43487 | 271.25 |
| 60 | 0.44879 | 279.58 |

We want to determine the properties of the refrigerant at the following temperatures (–12, 18, 32, 57):

1. Create a data file of the above data.
2. Create a MATLAB program that consists of a calling program and a self-written function.

**Calling program should do the following:**

1. Load the data file and create vectors Tt, vt, ut. Also create a vector, T, that includes the temperatures at which the properties are to be determined.
2. Call the self-written function using input variables of Tt, vt, ut, and T and output variables T, v, and u.
3. Print to the Command Window a table consisting of T, v, and u with table headings and units.
4. Plot vt versus Tt and on the same graph, plot v versus T as small x's.
5. Plot ut versus Tt and on the same graph, plot u versus T as small x's.

**Self-written function should do the following:**
Using the input variables and MATLAB's interp1 function, interpolate for the properties of the refrigerant at the temperatures, T, and return those values to the calling program.

**P4.6.** This project is a variation of Project P2.5. In that project, we discussed the pressure drop, $p_1-p_2$, in a circular pipe having a flow rate, Q [1]. We will repeat the governing equations involved in the process.

$$p_1 - p_2 = \frac{\rho V^2}{2} \frac{L}{D} f \quad \text{(P4.6a)}$$

where:
  $\rho$ is the fluid density (kg/m³)
  V is the average fluid velocity in the pipe (m/s)
  Q = AV is the volume flow rate in the pipe (m³/s)
  D is the pipe diameter (m)
  A is the pipe cross-sectional area = $(\pi D^2/4)$ (m²)
  L is the pipe length between points 1 and 2 (m)
  f is friction factor

The friction factor has been determined by experiment. For smooth pipes a formula that approximates the experimental data is [2]

$$f = (1.82 \log_{10} \mathrm{Re} - 1.64)^{-2} \quad \text{(P4.6b)}$$

where:

$$\mathrm{Re} = \frac{VD}{\upsilon} \quad \text{(Reynolds number)} \quad \text{(P4.6c)}$$

$$\upsilon \text{ is the fluid viscosity} \left(\frac{m^2}{s}\right) = \frac{\mu}{\rho} \quad \text{(P4.6d)}$$

Develop a MATLAB program that contains both a calling program and a function that determines the pressure drop, $p_1 - p_2$ versus the flow rate, $Q$. $Q$ is to vary from 0.001 to 0.02 m³/s in steps of 0.001 m³/s. Properties of $D$, $L$, $\rho$, $\upsilon$ and $Q$ are to be defined in the calling program. These values should be made as an input argument to the function. The output from the function is to be the pressure drop, $p_1 - p_2$, in the pipe and returned to the calling program. In the calling program, a table of $p_1 - p_2$ versus $Q$ is to be printed out to the Command Window, including table headings and units. Use the following values: $\rho = 1000$ kg/m³, $D = 0.16$ m, $L = 5$ m and $\upsilon = 1.2 \times 10^{-6}$ m²/s.

**P4.7.** This project is a modification of Project P3.5. That project involved the mass motion in a mass-spring-dashpot system. A sketch of such a system is shown in Figure 2.19. Disturbing the mass position, $y$, from its equilibrium position and releasing it with zero velocity, will result in the $y$ position varying with time, $t$.

As discussed in Project P3.5, the type of motion that the mass will have depends on the values of the system properties of $m$, $k$, and $c$, where $m$ is the mass, $k$ is the spring constant, and $c$ is the damping factor [3].

If, $(k/m) > (c/2m)^2$, then the mass motion will be damped oscillations and the governing equation describing the motion is

$$y = \exp\left(-\frac{c}{2m}t\right)\left\{A\cos\left(\sqrt{\frac{k}{m} - \left(\frac{c}{2m}\right)^2}\, t\right) + B\sin\left(\sqrt{\frac{k}{m} - \left(\frac{c}{2m}\right)^2}\, t\right)\right\} \quad \text{(P4.7a)}$$

The coefficients A and B are determined by initial conditions. Suppose we take the initial displacement to be 0.5 m. Then

$$A = 0.5 \text{ m}, \quad B = \frac{Ac}{2m\sqrt{\arg}}$$

where $\arg = k/m - (c/2m)^2$

## Self-Written Functions and MATLAB®'s fminbnd Function

If, $k/m < (c/2m)^2$, then the mass motion will be damped exponential motion and the governing equation will be

$$y = \exp\left(-\frac{c}{2m}t\right)\left\{A\exp\left(\sqrt{\left(\frac{c}{2m}\right)^2 - \frac{k}{m}}\, t\right) + B\exp\left(-\sqrt{\left(\frac{c}{2m}\right)^2 - \frac{k}{m}}\, t\right)\right\} \quad \text{(P4.7b)}$$

where $\exp(x) = e^x$. For this case,

$$A = \frac{0.5\left(\sqrt{-\text{arg}} + \dfrac{c}{2m}\right)}{2\sqrt{-\text{arg}}}, \quad B = A\,\frac{\sqrt{-\text{arg}} - \dfrac{c}{2m}}{\sqrt{-\text{arg}} + \dfrac{c}{2m}}$$

If $(c/2\,m)^2 = k/m$, then the system is *critically damped*. For this case, the solution is

$$y = (A + Bt)\exp\left(-\frac{c}{2m}t\right) \quad \text{(P4.7c)}$$

For this case,

$$A = 0.5 \text{ m}, \quad B = A\frac{c}{2m}$$

Construct a MATLAB program that consists of two parts: a calling program and a function. The calling program is to create four different vectors containing values of $m$, $k$, and $c$. The calling program is to use these vectors as an input to the function. The function is to determine which of the three equations to use in calculating vector $y$ as a function of vector $t$. Have the function return vectors $y$ and $t$ to the calling program where it is to create plots of $y$ versus $t$. Take $t = 0$ to 20 s, in steps of 0.1 s. Values of $m$, $k$, and $c$ for the four cases are listed below:

Case 1

$$m = 75 \text{ kg}, \quad k = 85\,\frac{N}{m}, \quad c = 200\,\frac{N\text{-}s}{m}$$

Case 2

$$m = 80 \text{ kg}, \quad k = 150\,\frac{N}{m}, \quad c = 40\,\frac{N\text{-}s}{m}$$

Case 3

$$m = 50 \text{ kg}, \ k = 50 \ \frac{\text{N}}{\text{m}}, \ c = 100 \frac{\text{N-s}}{\text{m}}$$

Case 4

$$m = 100 \text{ kg}, \ k = 200 \ \frac{\text{N}}{\text{m}}, \ c = 20 \frac{\text{N-s}}{\text{m}}$$

**P4.8.** Mathematician Joseph Fourier is credited with the theorem that any periodic waveform may be expressed as a summation of pure sines and cosines [4]. For example, the square wave of Figure P4.1a can be written as a sum of sine terms:

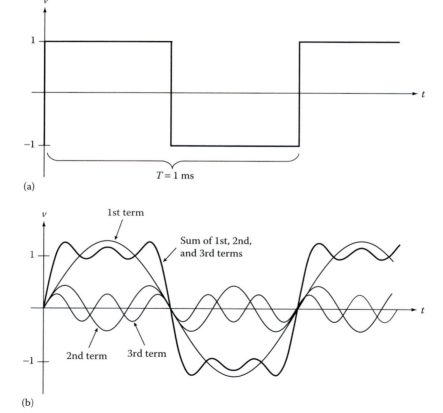

**FIGURE P4.1**
(a) Square wave and (b) using 1st, 2nd, and 3rd terms in the series and their sum.

## Self-Written Functions and MATLAB®'s fminbnd Function

$$v(t) = \frac{4}{\pi}\sin\frac{2\pi t}{T} + \frac{4}{3\pi}\sin\frac{6\pi t}{T} + \frac{4}{5\pi}\sin\frac{10\pi t}{T} + \cdots$$

$$= \sum_{\substack{k=1 \\ k\,\text{odd}}}^{\infty} \frac{4}{\pi k}\sin\frac{2\pi k t}{T}$$

(P4.8)

Figure P4.1b shows the first three terms of the series and their summation.

1. Write a MATLAB script that utilizes the self-written function sqwave(n,T,i) that takes the following input arguments:

    n is the number of terms of the Fourier series.

    T is the period of the square wave in seconds.

    i is the number samples points per period.

2. The function should return two arrays, t and v, each containing i elements, where:

    t is an array of i time points.

    v is an array of i computed values of the nth degree approximated square wave.

3. Run your sqwave(n,T,i) function and plot the results for the following arguments:

    T = 1 ms, i = 1001, n = 3, 10, 100

4. For n = 100, create a plot of v versus t.

**P4.9.** This project is a modification of Project P4.7. That project involved the mass motion in a mass-spring-dashpot system. A sketch of such a system is shown in Figure 2.19. Disturbing the mass position, y, from its equilibrium position and releasing it with zero velocity, will result in the y position varying with time, t. In this project we will consider the case $k/m > (c/2m)^2$, resulting in the mass motion being damped oscillations. The governing equation describing the motion is

$$y = \exp\left(-\frac{c}{2m}t\right)\left\{A\cos\left(\sqrt{\frac{k}{m} - \left(\frac{c}{2m}\right)^2}\,t\right) + B\sin\left(\sqrt{\frac{k}{m} - \left(\frac{c}{2m}\right)^2}\,t\right)\right\} \quad \text{(P4.7a)}$$

The coefficients A and B are determined by initial conditions. We will take the initial displacement to be 0.5 m and the initial velocity equal to zero. Then

$$A = 0.5 \text{ m},\ B = \frac{Ac}{2m\sqrt{\text{arg}}}$$

where:

$$\arg = \frac{k}{m} - \left(\frac{c}{2m}\right)^2$$

1. Plot $y$ versus $t$ for $t = 0$ to 20 s, in steps of 0.1 s.
2. Use MATLAB's `fminbnd` function to determine the minimum $y$ position for $t = 0$ to 5 s.

Take $m = 50$ kg, $k = 100(\text{N/m})$, $c = 20(\text{N}-\text{s/m})$.

**P4.10.** This Project is a variation of Exercise E2.4. A basketball player shoots the ball when he is 8 m from the center of the hoop, instead of the 6 m shown in Figure 2.20. The ball is released at a velocity, $V_o$, and makes angle $\vartheta_o = 45°$ with the horizontal. Using Newton's second law and the initial conditions and neglecting the drag on the basketball, we can determine the following equations for the $(x, y)$ position of the ball as a function of time, $t$:

$$x = V_o \cos(\vartheta_o) t \qquad (2.5)$$

$$y = V_o \sin(\vartheta_o) t + \frac{g}{2} t^2 \qquad (2.6)$$

Take the $(x, y)$ position of the center of the hoop to be $(x_f, y_f) = (8.0 \text{ m}, 3.048 \text{ m})$, $\vartheta_o = 45°$, $V_o = 9.5169$ m, and time of flight, $t_f = 1.1888$ s.

1. Create a plot of $y$ versus $t$ using $0 \leq t \leq t_f$ in steps of $(t_f/10)$.
2. Using MATLAB's `fminbnd` function determines the maximum height reached by the basketball in its flight to the basketball hoop.

---

## References

1. Bober, W., *Introduction to Numerical and Analytical Methods with MATLAB for Engineers and Scientist*, CRC Press, Boca Raton, FL, 2014.
2. Bober, W., The use of the Swamee-Jain formula in pipe network problems, *Journal of Pipelines*, 4, 315–317, 1984.
3. Thomson, W. T., *Theory of Vibration with Applications*, Prentice Hall, Englewood Cliffs, NJ, 1972.
4. Bober, W., Stevens, A., *Numerical and Analytical Methods with MATLAB for Electrical Engineers*, CRC Press, Boca Raton, FL, 2012.

# 5
# Working with Characters and Strings

## 5.1 Introduction

There may be a time that you might wish to create a matrix consisting of a string of characters and to print it out in a report. This chapter shows you how to do that. In MATLAB®, characters and strings usually need to be enclosed by single quotation marks.

**Example 5.1**

```
% Example_5_1.m
% This program demonstrates how to print out rows of character
% strings. This can be done by declaring a column vector where each
% element in the vector is a character string.
% Note that all row character strings must have the same number
% of columns and be enclosed by single quotation marks.
clear; clc;
% Assign a string column vector.
% Each row in the column vector must have the same number of columns.
parts=['Internal modem  '
       'Graphics adapter'
       'CD drive        '
       'DVD drive       '
       'Floppy drive    '
       'Hard disk drive '];
for i=1:6
    fprintf('%16s \n',parts(i,1:16));
end
```

**Program Results:**

```
Internal modem
Graphics adapter
CD drive
DVD drive
Floppy drive
Hard disk drive
>>
```

123

## Example 5.2

This example is an interactive program. The user has to input whether to print the string matrix to the screen or to a file.

```
% Example_5_2.m
% This example is a modification of Example 5.1.
% The program asks the user if he/she wishes to have the
% output go to the screen or to a file.
% This example also illustrates the use of the switch statement.
clear; clc;
parts=['Internal modem         '
       'External modem         '
       'Graphics circuit board'
       'CD drive               '
       'Hard disk drive        '];
fprintf('Choose whether to send the output to the\n');
fprintf('screen or to a file named output.txt. \n\n');
var=input('Enter S for screen or F for file (without quotes)\n','s');
switch(var)
    case 'S'
        for i=1:5
            fprintf('%22s \n',parts(i,1:22));
        end
    case 'F'
        fo=fopen('output.txt','w');
        for i=1:5
            fprintf(fo,'%22s \n',parts(i,1:22));
        end
        fclose(fo);
    otherwise
        fprintf('you did not enter an S or a F, try again \n');
        exit;
end
```
------------------------------------------------------------------

### Program Results:

```
Choose whether to send the output to the
screen or to a file named 'output.txt'.
Enter S for screen or F for file (without single quotes)
S
        Internal modem
        External modem
Graphics circuit board
              CD drive
       Hard disk drive
>>
```
------------------------------------------------------------------

## Example 5.3

In this interactive example, we illustrate the use of the `if-elseif` ladder to establish a letter grade when the user enters a numerical score.

```
% Example_5_3.m
% This example uses the if-elseif ladder.
% The program determines a letter grade depending on the score the user
% enters from the keyboard.
clear; clc;
gradearray=['A'; 'B'; 'C'; 'D'; 'F'];
score=input('Enter your test score: \n');
fprintf('score is: %i \n',score);
    if score > 100
        fprintf('error: score is out of range. Rerun program \n');
        break;
    elseif score >= 90 && score <= 100
        grade=gradearray(1);
    elseif score >= 80 && score < 90
        grade=gradearray(2);
    elseif score >= 70 && score < 80
        grade=gradearray(3);
    elseif score >= 60 && score < 70
        grade=gradearray(4);
    elseif score < 60
        grade=gradearray(5);
    end
fprintf('grade is: %c \n',grade);
```

**Program Results:**
```
Enter your test score: 76
score is: 76
grade is: C
>>
```

## Example 5.4

This example is a modification of Example 5.3. In this interactive example a `for` loop is used to establish the interval containing the grade.

```
% Example_5_4.m
% The program determines a letter grade depending on the score the user
% enters from the keyboard.
% This version uses a loop to determine the correct interval of
```

```
% interest.  For a large number of intervals, this method is more
% efficient (fewer statements) than the method in Example 5.3
clear; clc;
gradearray=['A'; 'B'; 'C'; 'D'; 'F'];
sarray=[100 90 80 70 60 0];
score=input('Enter your test score: \n');
fprintf('score is: %i \n',score);
% The following 3 statements are needed for the case when score = 100.
if score == 100
    grade=gradearray(1);
else
    for i=1:5
        if score >= sarray(i+1) && score < sarray(i)
            grade=gradearray(i);
            break;
        end
    end
end
fprintf('grade is: %c \n',grade);
```
---

**Program Results:**
```
Enter your test score: 82
score is: 82
grade is: B
>>
```
---

### Example 5.5

This example combines the use of a string matrix and the establishment of a grade.

```
% Example_5_5.m
% This program determines the letter grades of several students.
% Student's names and their test scores are entered in the program.
% Student names are not connected to real people.
% This example uses nested 'for' loops and an 'if' statement
% to determine the correct letter grade for each student.
clear; clc;
gradearray=['A'; 'B'; 'C'; 'D'; 'F'];
sarray=[100 90 80 70 60 0];
Lname=['Smith     '
       'Lambert   '
       'Kurtz     '
       'Jones     '
       'Hutchinson'
       'Blake     '];
Fname=['Joe  '
       'Jane '
       'Howard'
       'Mary '
       'Peter'
       'Henry'];
score=[84; 86; 67; 92; 81; 75];
```

## Working with Characters and Strings

```
avg_score=mean(score);
fprintf('The group average numerical grade is:%4.1f \n',avg_score);
% Index j selects the student and index i selects the letter grade.
% The score = 100 is treated separately.
for j=1:6
    if score(j) == 100
        grade(j)=gradearray(1);
    else
        for i=1:5
            if score(j) >= sarray(i+1) && score(j) < sarray(i)
                grade(j)=gradearray(i);
            end
        end
    end
end
fprintf('Last name     First name     grade  \n');
fprintf('-------------------------------------\n');
for j=1:6
    fprintf('%12s      %10s       %c \n',...
            Lname(j,1:12), Fname(j,1:10), grade(j));
end
```
---

**Program Results:**

```
The group average numerical grade is:80.8
Last name     First name    grade
------------------------------------
Smith         Joe           B
Lambert       Jane          B
Kurtz         Howard        D
Jones         Mary          A
Hutchinson    Peter         B
Blake         Henry         C
>>
```
---

## 5.2 MATLAB's `textscan` Function

There may be occasions when you wish to enter information into a program from a data file that contains both numerical and text data. MATLAB's `textscan` function is best suited for this operation.

Syntax:

```
C = textscan(fo, format)
```

The function will read data from an open text file identified by `fo` into a cell array `C`. The format is of the form `%f`, `%d`, `%c`, `%s`, and so on. The number of format specifiers determines the number of cells in the cell array `C`. Each cell will contain the number of lines contained in the data file and be of the type specified by the format statement. String specifiers also include %q, which is a string enclosed by double quotation marks. In the textscan statement, the format for a string of n characters is %nc, but in the print statement, the format for a string of n characters is %ns.

**NOTE:** To reference the contents of a cell, enclose the cell number with { }. See the following example:

If you wish to read in N lines from the open data file, use

```
C = textscan('fo', format,N)
```

### Example 5.6

Loads mixed text and numerical data from a file

```
% Example_5_6.m
% Load the product data from inv4.txt into the arrays 'cat_nu', 'desc',
% 'cost', and print the results to the screen.
clear; clc;
fo=fopen('inv4.txt'); % Note: inv4.txt is defined below.
C = textscan(fo,'%d %14c %f',5);
% Contents of cell block C contains 5 rows and 3 columns
cat_nu = C{1};
desc = C{2};
cost = C{3};
fclose(fo);
fprintf('catalog #    description \t\t   cost  \n');
fprintf('------------------------------------------\n');
for i=1:5
    fprintf('%5i \t %14s \t %6.2f \n\n',...
            cat_nu(i),desc(i,1:14),cost(i));
end
------------------------------------------------------------------
% inv4.txt file (do not include this line in your data file)
1001      hammer            2.58
1002      plier             1.20
1003      screwdriver       1.56
1004      soldering iron    3.70
1005      wrench            2.60
------------------------------------------------------------------
```

### Program Results:

```
catalog #    description       cost
--------------------------------------
1001       hammer             2.58
1002       plier              1.20
1003       screwdriver        1.56
1004       soldering iron     3.70
1005       wrench             2.60
>>
------------------------------------------------------------------
```

# Working with Characters and Strings

> **REVIEW 5.1**
>
> 1. Suppose you wish to assign a column vector consisting of string elements, what are the conditions that need to be followed in setting up this column vector?
> 2. Suppose that you had a data file that contains both numerical and text data, what command would you use to read in the data into your program.
> 3. When the command used in reading in the data type described in item 2, what object type does the data go into?
> 4. To assign variable names to items in the object which of the following three symbols would you use: (), [], {}?

**Projects**

**P5.1.** Create a MATLAB program that

1. Contains a string array that lists the names of five courses that you have taken recently.
2. Contains a vector of the numerical grades received in each of the five courses.
3. Prints to the screen the names of the five courses and their grades.
4. Prints to the screen the average grade received in the five courses.

**P5.2.** You are a student adviser and you wish to have the user (student) tell you the grades that he/she received on several courses. You are to use MATLAB's `input` command to ask the student to enter the grade that he received in the course.
Create a MATLAB program that

1. Creates a string matrix of five courses.
2. Use a `for` loop and MATLAB's `input` command asking the student to enter the grade that he/she received in the course listed on the screen. List the courses one at a time.
3. Print to the screen a summary of the results.

NOTE: Allowable grades are only letters (A, B, C, D, F, I).

**P5.3.** Create a MATLAB program that creates a vector listing all the days of the week, that is, Monday, Tuesday, and so on. Create a second vector that describes two activities that you wish to do on each day of the week, such as

*Attend calculus class and have dinner with girl or boy friend*, and so on. Do this for each day of the week. Use an input statement to have the user enter the day of the week. Based on that entry and using a Switch Statement print out the activities of that day listed in the second vector.

**P5.4.** Bob's Hardware Store wishes to create an online program to sell inventory items in its store. You are to create an interactive MATLAB program for this purpose. The program is to contain a data file, a main program and a billing function.

Data file:

> The data file is to contain a catalog number and a description of the inventory items for sale, their costs, and the quantity available for sale. The list should contain at least 10 items. The data file is to be loaded into the main script.

Main script:

1. The script is to print to the screen the items for sale, including their catalog numbers and descriptions, their cost, and the quantity available for purchase.
2. The script is to ask the user if he/she wishes to make a purchase. If no, exit the program. If yes, the script is to ask the user for his/her first and last name. It should then ask the user the catalog number of the item he/she wishes to purchase and the quantity.
3. The script is to continue asking the user if he/she wishes to make another purchase. If the response is no, call the billing function and exit the program. If the response is yes, print to the screen the list of items, their cost, and the updated items available for purchase. Then, the script is to ask the user the catalog number of the item he/she wishes to purchase and the quantity.
4. Billing function:

    Input arguments:

    First and last names of the purchaser, the number of items purchased, the catalog numbers, their costs, the quantities of the items purchased.

    Output arguments:

    None, use open and closed brackets.

    Use a global statement for the description of the items purchased (both in the calling program and in the billing function). The bill should contain the name and address of the store, the customer's first and last names, bill headings, the catalog numbers of all the items purchased, their unit prices, the total price for each item purchased, and finally the total price of all the items purchased.

    Clear the screen, then print the bill to the screen.

# 6

# Roots of Algebraic and Transcendental Equations

## 6.1 Introduction

In the analysis of various engineering problems, we are often faced with a need to find roots of equations whose solution is not easily found analytically. Given a function $f(x)$, the roots of the function are the values of $x$ that makes $f(x) = 0$. For example, the equation

$$f(x) = ax^2 + bx + c = 0 \qquad (6.1)$$

where $a$, $b$, and $c$ are constants, is an equation that we are all familiar with. The values of $x$ that satisfy the equation are the roots of $f(x)$. We even have a formula for the roots, which are

$$x = \frac{-b \pm \sqrt{b^2 - 4ac}}{2a} \qquad (6.2)$$

We see that there are two roots, $x1$ and $x2$, where

$$x1 = \frac{-b + \sqrt{b^2 - 4ac}}{2a}, \quad x2 = \frac{-b - \sqrt{b^2 - 4ac}}{2a} \qquad (6.3)$$

More complicated examples include $n$th degree polynomials and transcendental equations containing trigonometric, exponential, or logarithm functions. In this chapter, we discuss the search method for obtaining a small interval in which a root lies. We then discuss MATLAB®'s `fzero` and `roots` functions, which may be used to obtain a more accurate value for the roots of type of equations just stated.

## 6.2 Search Method

In the search method, we seek a small interval that contains a real root. This only gives an approximate value for the real root. Once an interval in which a real root lies has been established, several different methods, including the Bisection method, Newton–Raphson method, and MATLAB's fzero and roots functions, can be used to obtain a more accurate value for the real root. In this book, we will give a brief discussion of the Bisection method, but emphasize MATLAB's fzero and roots functions.

The search method is especially useful if there is more than one real root. The equation whose roots are to be determined should be put into the following standard form:

$$f(x) = 0 \quad (6.4)$$

We proceed as follows: first we subdivide the $x$ domain into $N$ equal subdivisions of width $\Delta x$, giving

$$x_1, x_2, x_3, \ldots x_{N+1} \quad \text{with} \quad x_{i+1} = x_i + \Delta x$$

Then, determine where $f(x)$ changes sign (see Figure 6.1).

This occurs when the signs of two consecutive values of $f(x)$ are different, that is,

$$f(x_i)f(x_{i+1}) < 0$$

The sign change usually indicates that a real root has been passed. However, it may also indicate a discontinuity in the function. (Example: $\tan x$ is discontinuous at $x = \pi/2$.)

A brief description of the Bisection method follows:

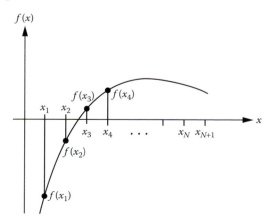

**FIGURE 6.1**
The root of $f(x)$ lies between $x_2$ and $x_3$.

## 6.3 Bisection Method

Suppose it has been established by the search method, that a root lies between $x_i$ and $x_{i+1}$. The concept in the bisection method is to cut the interval containing the root in half, determine which half contains the root, cut that interval in half, determine which half contains the root, and continue the process until the interval containing the root is sufficiently small, so that any point within the last interval is a very good approximation for the root. A more detailed description follows: Let $x_{i+\frac{1}{2}}$ be the midpoint position of the first cut, then $x_{i+\frac{1}{2}} = x_i + (\Delta x/2)$ (see Figure 6.2). Now compute $f(x_i)f(x_{i+\frac{1}{2}})$:

Case 1: If $f(x_i)f(x_{i+\frac{1}{2}}) < 0$, then the root lies between $x_i$ and $x_{i+\frac{1}{2}}$
Case 2: If $f(x_i)f(x_{i+\frac{1}{2}}) > 0$, then the root lies between $x_{i+\frac{1}{2}}$ and $x_{i+1}$
Case 3: If $f(x_i)f(x_{i+\frac{1}{2}}) = 0$, then $x_i$ or $x_{i+\frac{1}{2}}$ is a real root

For cases 1 and 2, select the interval containing the root and repeat the process. Continue repeating the process, say $r$ times, then $(\Delta x)_f = \Delta x/2^r$, where $\Delta x$ is the initial size of the interval containing the root before the start of the bisection process ($\Delta x = x_{i+1} - x_i$) and $(\Delta x)_f$ is the size of the interval containing the root after $r$ bisections. If $(\Delta x)_f$ is sufficiently small, then a very good approximation for the root is anywhere within the last bisected interval, say the midpoint of the interval. For example, in 20 bisections,

$$(\Delta x)_f = \frac{\Delta x}{2^{20}} \approx \Delta x \times 1.0 \times 10^{-6}$$

---

MATLAB has built-in functions to determine the roots of a function of one variable, such as a transcendental equation or an $n$th degree polynomial. The `fzero` function is used for transcendental equations and will determine the real roots of the equations. The `roots` function is used for polynomial equations and will return both the real and imaginary roots. First we will discuss the `fzero` function.

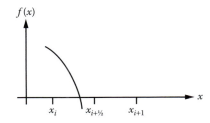

**FIGURE 6.2**
Bisecting the interval containing the root.

## 6.4 MATLAB's fzero Function

The fzero function is for a single variable nonlinear function whose root we wish to determine. The syntax for the fzero function is

$$X = \text{fzero}(\text{FUN}, \text{X0}) \quad (6.5)$$

where FUN is a function handle to the function whose root is to be determined, X0 is a scalar and represents an initial guess for the root, and X is the root determined by the fzero function. FUN may be a separate .m file or an anonymous function. You may also enter the function directly into the fzero function using the function creator, @, as shown in Example 6.2. To get a good value for X0, consider plotting the function and noting where the function crosses the *x*-axis.

**Example 6.1**

Given the equation

$$f(v) = \frac{3.3}{v - 0.03} - \frac{0.325}{v(v + 0.03)} - 2 \quad (6.6)$$

Determine the root of *f*(v).

```
% Example_6_1.m
% Simple use of fzero function
% Find the root of f(v)=3.3/(v-0.03)-0.325/(v(v+0.03))-2
% Guess that the root lies between 1 and 2.
% Function name is fv and the function is a .m file.
clear; clc;
v=1.0:0.1:2.0;
for j=1:length(v)
    f(j)=fv(v(j));
end
plot(v,f), xlabel('v'), ylabel('f'), title('f vs. v'), grid;
root=fzero('fv',1.6);
% Note: we could also have used root=fzero(@fv,1.6)
fprintf('root=%6.4f \n',root);
froot=fv(root);
fprintf('froot=%12.4e \n',froot);
%--------------------------------------------------------------
% This function works with Example 6.1
function f=fv(v)
f=3.3/(v-0.03)-0.325/(v*(v+0.03))-2;
%--------------------------------------------------------------
```

**Program Results:**

```
root= 1.5810
froot= 0.0000e+00
>>
```
See Figure 6.3.
--------------------------------------------------------------

# Roots of Algebraic and Transcendental Equations

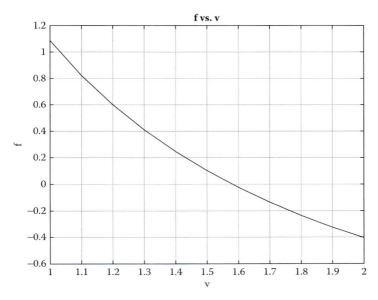

**FIGURE 6.3**
Plot of f versus v.

### Example 6.2

In the next two examples, we will determine the root of Equation 6.5 by placing an anonymous function directly into the `fzero` function and also as by placing an anonymous function in the main script. In both cases no separate .m file is used.

```
% Example_6_2.m
% Anonymous function applied directly into the fzero function
% Find the root of f(v)=3.3/(v-0.03)-0.325/(v(v+0.03))-2
% Guess that the root lies between 1 and 2.
clear; clc;
root=fzero(@(v) (3.3/(v-0.03)-0.325/(v*(v+0.03))-2),1.6);
fprintf('root=%6.4f \n',root);
froot=fv(root);
fprintf('froot=%12.4e \n',froot);
-------------------------------------------------------------
```

**Program Results:**
```
root= 1.5810
froot= 0.0000e+00
>>
-------------------------------------------------------------

% Example_6_3.m
% Anonymous function used with the fzero function.
% Find the root of f(v)=3.3/(v-0.03)-0.325/(v(v+0.03))-2
```

```
% Guess that the root lies between 1 and 2.
clear; clc;
fv=@(v) (3.3/(v-0.03)-0.325/(v*(v+0.03))-2);
v=1.0:0.1:2.0;
for j=1:length(v)
    f(j)=fv(v(j));
end
plot(v,f), xlabel('v'), ylabel('f'), title('f vs. v'), grid;
% Note: When using an anonymous function for FUN do not use
% either the @ sign or enclose the name of the anonymous function
% with single quotation marks.
root=fzero(fv,1.6);
fprintf('root=%6.4f \n',root);
froot=fv(root);
fprintf('froot=%12.4e \n',froot);
```

**Program Results:**

The results are exactly the same that was obtained in Example 6.1.

In some instances, we would like to find the zero of a function of two arguments, say X and P, where P is a parameter and is fixed. In order to solve with fzero, P must be defined in the calling program. For example, suppose myfun is defined in a .m file as a function of two arguments:

```
function f = myfun(X,P)
f = cos(P*X);
```

The fzero statement would need to be invoked as follows:

```
P = 1000;
root = fzero(@(X) myfun(X,P),X0)
```

where root is the zero of function myfun when P=1000. Note that P needs to be defined *before* the fzero function is called.

An alternative to adding parameter P as an argument in myfun is to use MATLAB's global statement. The parameter P should be defined in the calling program and be listed as a variable in the global statement. The global statement needs to be used in both the calling program and in the function myfun and be exactly the same in both scripts. See the following example.

**Example**

Calling program:

```
global P;
P=1000;
Xo=10.0;
root = fzero(@myfun,Xo);
```

```
% The file myfun.m:
function f = myfun(x)
global P;
f = cos(P*x);
```
---

Now, let us consider the case when there is more than one root in the function under consideration. First, it is best to use the search method to obtain small intervals in which the roots lie. For each obtained interval define the argument X0 in the fzero command (see Equation 6.5) as a vector of length two; that is, X0(1) is the X position at the beginning of one of the found intervals and X0(2) is the X position at the end of that interval. This should result in the sign of FUN(X0(1)) to differ from the sign of FUN(X0(2)). If that is not the case, MATLAB will return an error message. The following example illustrates this concept:

**Example 6.4**

The position, $y$, of a mass in a mass-spring-dashpot system (see Figure 2.19) that is underdamped is given by

$$y = \exp\left(-\frac{c}{2m}t\right)\left\{A\cos\left(\sqrt{\frac{k}{m}-\left(\frac{c}{2m}\right)^2}\,t\right)+B\sin\left(\sqrt{\frac{k}{m}-\left(\frac{c}{2m}\right)^2}\,t\right)\right\} \qquad (6.7)$$

Determine the number of roots, their values and $y$ values at the obtained roots for $0 \le t \le 10$ s. Take
$m = 25.0$ kg, $c = 5.0$ N-s/m, $k = 200.0$ N/m, $A = 0.2$ m, and

$$B = \frac{c}{2m} \times \frac{A}{\sqrt{k/m-(c/2m)^2}}$$

```
% Example_6_4.m
% This program determines the number of roots and their values
% in the mass-spring-dashpot system in the time span 0<=t<=10 s.
% The governing equation for the displacement, y(t), of the
% under-damped vibration problem is:
% y(t)=exp(-c*t/2/m)(A*cos(arg1*t)+B*sin(arg1*t)), where
% arg1=sqrt(k/m-(c/2/m)^2)
% k=200 N/m, m=25kg, c=5 N-s/m
% A=0.2m, B=c/(2*m)*A/arg1.
% The values for constants A and B represent an initial
% displacement of the mass from its equilibrium position
% at zero velocity.
% A global statement in used to bring the constants, k, m, c,
% A, B and arg1 into the function fun_spring.
clear; clc;
global m k c A B arg1;
m=25; k=200; c=5; A=0.2;
arg1=sqrt(k/m-(c/2/m)^2);
B=c/2/m*A/arg1;
```

```
% ir is the root number
ir=0;
t=0:0.1:10;
for i=1:length(t)
    y(i)=fun_spring(t(i));
end
plot(t,y), xlabel('t(s)'), ylabel('y(m)'),
title('mass displacement, y, vs. t'), grid;
for i=1:length(t)-1
    if y(i)*y(i+1) <= 0.0
        ir=ir+1;
        tr(1)=t(i);
        tr(2)=t(i+1);
        root(ir)=fzero('fun_spring',tr);
        y(ir)=fun_spring(root(ir));
    end
end
if ir ~= 0
    fprintf('root #     root(s)       y(root)  (m)   \n');
    fprintf('------------------------------------------\n');
    for j=1:ir
        fprintf('%3i         %10.6f     %12.4e \n',j,root(j),y(j));
    end
else
    fprintf('\n\n No roots lie within 0 <= t <= 20 s');
end
-----------------------------------------------------------------------
% fun_spring.m
% This function is used in Example 6.4
% The function determines the spring position, y(t), as a
% function of time
function y=fun_spring(t)
global m k c A B arg1;
y=exp(-c*t/2/m)*(A*cos(arg1*t)+B*sin(arg1*t));
-----------------------------------------------------------------------
```

**NOTE:** We could have avoided the use of the global statement by defining the values of m, k, c, A, B, and arg1 in the function instead of the main program. We chose to use the `global` statement to illustrate its use.

**Program Results:**

```
root #      root(s)       y(root)(m)
----------------------------------------
   1        0.568218      -1.3111e-17
   2        1.679634      -1.2099e-16
   3        2.791049      -7.2174e-18
   4        3.902465       1.0979e-16
   5        5.013881      -2.9893e-16
   6        6.125296       6.1582e-17
   7        7.236712      -4.2906e-17
   8        8.348127       2.8230e-17
   9        9.459543      -1.5493e-17
>>
```
See Figure 6.4.
-----------------------------------------------------------------------

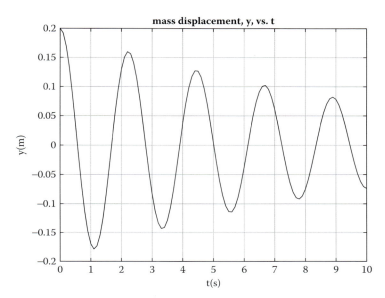

**FIGURE 6.4**
Plot of mass motion in a mass-spring-dashpot system.

## 6.5 MATLAB's roots Function

MATLAB has a function named roots that obtains the roots of a polynomial. The function determines both real and imaginary roots of the specified polynomial.

The syntax for the function is

V = roots(C) where C is a vector specifying the coefficients of the polynomial and V is the roots. If C has n + 1 components, the polynomial is

$$C(1)x^n + C(2)x^{n-1} + ... C(n)x + C(n+1) = 0$$

Thus, to find the roots of the polynomial $ax^4 + bx^3 + cx^2 + dx + e = 0$, run roots([a b c d e]). The roots function will give both real and imaginary roots of the polynomial.

Some additional useful MATLAB functions are

poly(V) finds the coefficients of the polynomial whose roots are V.
real(V) gives the real part of V.
imag(V) gives the imaginary part of V.

## Example 6.5

In this example, MATLAB's roots function is used to find the roots of a polynomial.

```
% Example_6_5.m
% This program determines the roots of two different polynomials
% using MATLAB's 'roots' function.
clear; clc;
% The first polynomial is: f=x^3-5.7*x^2-35.1*x+85.176. The
% roots of this polynomial are all real.
% Define coefficients of first polynomial (real roots)
fprintf('The coefficients of the first polynomial are: \n');
C=[1.0 -5.7 -35.1 85.176]
fprintf('The roots are: \n');
V=roots(C)
fprintf('Polynomial coefficients determined from poly(V) are:\n');
C_recalc=poly(V)
fprintf('----------------------------------------\n');
% The second polynomial is: f=x^3-9*x^2+23*x-65. The roots of
% this polynomial are both real and complex. Complex roots must
% be complex conjugates.
% Define the coefficients of second polynomial
fprintf('The coefficients of the second polynomial are \n');
D=[1.0 -9.0 23.0 -65.0]
fprintf('The roots are: \n');
W=roots(D)
fprintf('The real and imaginary parts of the roots are:\n');
re=real(W)
im=imag(W)
fprintf('Polynomial coefficients determined from poly(W) are:\n');
W_recalc = poly(W)
```
----------------------------------------

### Program Results:

```
The coefficients of the first polynomial are:
C =
    1.0000  -5.7000  -35.1000  85.1760
The roots are:
V =
    8.6247
   -4.9285
    2.0038
Polynomial coefficients determined from poly(V) are:
C_recalc =
    1.0000  -5.7000  -35.1000  85.1760
----------------------------------------
The coefficients of the second polynomial are
D =
    1    -9    23    -65
The roots are:
W =
    7.0449 + 0.0000i
```

# Roots of Algebraic and Transcendental Equations

```
    0.9775 + 2.8759i
    0.9775 - 2.8759i
The real and imaginary parts of the roots are:
re =
    7.0449
    0.9775
    0.9775
im =
         0
    2.8759
   -2.8759
Polynomial coefficients determined from poly(W) are:
W_recalc =
    1.0000  -9.0000  23.0000  -65.0000
>>
```

---

## REVIEW 6.1

1. What is meant by the term root of function $f(x) = 0$?
2. What is the objective in the search method for determining a root of the equation $f(x) = 0$?
3. What is the name of the MATLAB function for determining the roots of a transcendental equation of the form $f(x) = 0$?
4. In MATLAB's function for determining the roots of a transcendental equation, how does one define the function whose roots are to be determined?
5. If you suspect that there is more than one real root, what method should be used in combination with the MATLAB's fzero function to obtain the roots?
6. If you are using the search method in combination with the fzero function, what can you say about the second argument in the fzero function?
7. What is the purpose of the global statement?
8. If the function $f(x)$ is a polynomial, what MATLAB function should you use to obtain its roots?

## Projects

**P6.1.** This project is a variation of Project P2.1. In that project a tennis player on serve places the tennis ball close to the outside line of the service box when the ball hits the ground (see Figure P6.1a and b). The horizontal distance from the point where the ball leaves the racket to where the ball hits the ground is 18.925 m. The horizontal distance from the point where the ball leaves the racket to where the net is located is 12.509 m. The vertical distance above the ground when the ball leaves the racket is $y_o = 2.438$ m. The speed of the ball as it leaves the racket is 58.0 m/s. Determine the angle, $\vartheta_o$, that the ball makes with the horizontal on leaving the players racket that would result in the ball hitting the ground at the position stated. Neglecting drag, the governing equations describing the motion are

$$x = V_o \cos(\vartheta_o) t \quad \text{(P6.1a)}$$

$$y = -\frac{g}{2} t^2 - V_o \sin(\vartheta_o) t + y_o \quad \text{(P6.1b)}$$

Hint: Let $(x_f, y_f)$ be the position and $t_f$ be the time when the ball hits the ground. Substitute these values in Equations P6.1a and P6.2a, then solve for $t_f$ in the modified Equation P6.1a and substitute the obtained expression into the modified Equation P6.1b. This gives a transcendental

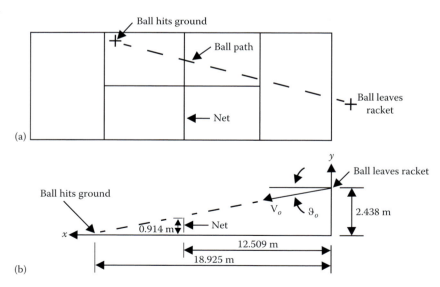

**FIGURE P6.1**
(a) Tennis court layout. (b) Path of tennis ball after it leaves the racket.

equation for $\vartheta_o$ where $\vartheta_o$ is the only unknown. Note $y_f = 0$. Write a program in MATLAB that

1. Plots $f(\vartheta_o)$ versus $\vartheta_o$.
2. Uses MATLAB's `fzero` function to determine $\vartheta_o$.
3. Determine whether the obtained value for $\vartheta_o$ will result in the ball clearing the net.
4. Find the time, $t_f$, it takes for the ball to hit the ground after it leaves the racket.
5. Using circles, plot the x-y position of the ball for $0 \le t \le t_f$ in steps of $t_f/10$.
6. Determine the height of the tennis ball when it reaches the x position of the net. Does the tennis ball clear the net?

Assume that $\vartheta_o$ can range anywhere from 0° to 8° in steps of 0.1°.

**P6.2.** This Project is a variation of Exercise E2.4. A basketball player shoots the ball when he is 6 m from the center of the hoop as shown Figure 2.20. The ball is released at a velocity, $V_o = 8.71$ m/s, and at an angle $\vartheta_o$ with the horizontal. Using Newton's second law and the initial conditions and neglecting the drag on the basketball, we can determine the following equations for the (x, y) position of the ball as a function of time, t:

$$x = V_o \cos(\vartheta_o) t \qquad \text{(P6.2a)}$$

$$y = \frac{g}{2} t^2 + V_o \sin(\vartheta_o) t + y_o \qquad \text{(P6.2b)}$$

Take the (x, y) position of the center of the hoop to be $(x_f, y_f) = (6.0 \text{ m}, 3.048 \text{ m})$ and $y_o = 1.98$ m.

1. Determine the angle $\vartheta_o$ that will result in the ball reaching the center of the hoop at time $t_f$.
2. Determine the time, $t_f$, that it takes for the ball to reach the center of the hoop. Time, t, equals zero when the ball leaves the player's hands.
3. Create a table consisting of t, x, y for $0 \le t \le t_f$ in steps of $t_f/10$. Carry variables to four decimal places. Print the table to an output file, including $t_f$ and $\vartheta_o$.
4. Create a plot of y versus x.

**P6.3.** The equation of state for a substance is a relationship between pressure, p, temperature, T, and specific volume, $\bar{v}$. Many gases at low pressures and moderate temperatures behave approximately as an ideal gas. The ideal

**TABLE P6.1**
Values of $a$, $b$ and, $\bar{R}$ for Carbon Dioxide in Redlich–Kwong's Equation of State

| Gas | $a\left(\dfrac{\text{N-m}^4\text{-K}^{1/2}}{\text{kmol}^2}\right)$ | $b\left(\dfrac{\text{m}^3}{\text{kmol}}\right)$ | $\bar{R}\left(\dfrac{\text{N m}}{\text{K-kmol}}\right)$ |
|---|---|---|---|
| Carbon dioxide | $65.43 \times 10^5$ | 0.02963 | 8314 |

*Source:* Moran, M. J. and Shapiro, H. N., *Fundamentals of Thermodynamics*, John Wiley & Sons, Hoboken, NJ, 2004.

gas equation of state with $p$ in N/m², $\bar{v}$ in m³/kmol, $T$ in K, and $\bar{R}$ in (N-m)/(K-kmol) is

$$p = \frac{\bar{R}T}{\bar{v}} \tag{P6.3a}$$

where $\bar{R}$ is the universal gas constant and is the same for all gasses. As temperature decreases and pressure increases, gas behavior deviates from ideal gas behavior. The Redlich–Kwong's equation of state is often used to approximate non-ideal gas behavior. Redlich–Kwong's equation of state is [1]

$$p = \frac{\bar{R}T}{\bar{v}-b} - \frac{a}{\bar{v}(\bar{v}+b)T^{1/2}} \tag{P6.3b}$$

or

$$f(\bar{v}) = \frac{\bar{R}T}{\bar{v}-b} - \frac{a}{\bar{v}(\bar{v}+b)T^{1/2}} - p = 0 \tag{P6.3c}$$

The values for $\bar{R}$, $a$, and $b$ for carbon dioxide is tabulated in Table P6.1.

We wish to determine the % error in the specific volume by using the ideal gas relationship while assuming that Redlich–Kwong's equation of state is the correct equation of state for carbon dioxide. Vary the temperature from 350–700 K in steps of 50 K, while holding the pressure constant at $1.0132 \times 10^7$ N/m² (100 atm). Using the specified temperatures and pressure determine the specific volumes, $\bar{v}$, by both the ideal gas equation and the Redlich–Kwong's equation and determine the % error in the specific volume resulting from the use of the ideal gas equation. Take the % error in the specific volume to be

$$\% \text{ error} = \frac{|\bar{v}_{\text{ideal gas}} - \bar{v}_{\text{Redlich–Kwong}}|}{\bar{v}_{\text{Redlich–Kwong}}} \times 100 \tag{P6.3d}$$

Write a MATLAB program utilizing the `fzero` function to calculate the specific volume by Redlich–Kwong's equation. Use the value of $\bar{v}$ obtained from the ideal gas law as your guess for the root in MATLAB's `fzero` function. Construct a table as shown in Table P6.2.

## TABLE P6.2
$\bar{v}$ Determined by Redlich–Kwong Equation and by Ideal Gas Law for Carbon Dioxide

| T(K) | Ideal Gas $\bar{v}$ (m³/kmol) | Redlich–Kwong Equation $\bar{v}$ (m³/kmol) | % Error in $\bar{v}$ |
|---|---|---|---|
| 350 | --------- | --------- | --------- |
| 400 | --------- | --------- | --------- |
| ----- | --------- | --------- | --------- |
| ----- | --------- | --------- | --------- |
| 700 | --------- | --------- | --------- |

**P6.4.** Determine the first root of the voltage, v(t), of the underdamped parallel RLC circuit described in Project P2.7. The governing equation for v(t) is

$$v(t) = \exp\left(-\frac{1}{2RC}t\right)\left\{A\cos\left(\sqrt{\frac{1}{LC} - \left(\frac{1}{2RC}\right)^2}\, t\right) + B\sin\left(\sqrt{\frac{1}{LC} - \left(\frac{1}{2RC}\right)^2}\, t\right)\right\} \quad \text{(P6.4)}$$

Use MATLAB's `fzero` function to find the first root. Print this value to the screen. Also plot v versus t. Assume the following parameters:

$$R = 100\ \Omega,\quad L = 10^{-3}\ \text{H},\quad C = 10^{-6}\ \text{F},\quad A = 6.0\ \text{V},\quad B = -9.0\ \text{V}$$

$0 \le t \le 5.0 \times 10^{-4}$ s in steps of $1.0 \times 10^{-5}$ s.

**P6.5.** Repeat Project 6.3, but replace the Redlich–Kwong's Equation with Van der Waals' Equation [1]. In addition, do the Project for all three gasses listed in Table P6.3 by the use of a `for` loop. In your program, use an `if-elseif` ladder within the `for` loop to select the proper constants for the gas. Van der Waals' equation of state is

$$p = \frac{\bar{R}T}{\bar{v} - b} - \frac{a}{\bar{v}^2} \quad \text{(P6.5)}$$

The constants a and b are tabulated in Table P6.3.

## TABLE P6.3
Van der Waals' Constants

| Gas # | Gas | $a\left(\dfrac{N\,m^4}{kmol^2}\right)$ | $b\left(\dfrac{m^3}{kmol}\right)$ | $\bar{R}\left(\dfrac{N\,m}{K\,kmol}\right)$ |
|---|---|---|---|---|
| 1 | Air | $1.368 \times 10^5$ | 0.0367 | 8314 |
| 2 | Oxygen | $1.369 \times 10^5$ | 0.0317 | 8314 |
| 3 | Carbon dioxide | $3.647 \times 10^5$ | 0.0428 | 8314 |

*Source:* Moran, M. J. and Shapiro, H. N., *Fundamentals of Thermodynamics*, John Wiley & Sons, Hoboken, NJ, 2004.

**P6.6.** The temperature distribution of a thick flat plate, initially at a uniform temperature, $T_0$, and which is suddenly immersed in a huge bath at a temperature $T_\infty$, is given by (see Figure P6.2a on the next page)

$$T(x,t) = T_\infty + 2(T_0 - T_\infty) \sum_{n=1}^{\infty} \frac{\sin(\delta_n)\cos\left(\delta_n \frac{x}{L}\right) e^{-a\delta_n^2 t/L^2}}{\cos(\delta_n)\sin(\delta_n) + \delta_n} \quad \text{(P6.6a)}$$

where:
 $L = 1/2$ of the plate thickness
 $a$ is the the thermal diffusivity of the plate material
 $\delta_n$ are the roots of the equation:

$$F(\delta) = \tan\delta - \frac{hL}{k\delta} = 0 \quad \text{(P6.6b)}$$

where:
 $h$ is the convective heat transfer coefficient for the bath
 $k$ is the thermal conductivity of the plate material

There are an infinite number of roots to Equation P6.5b. This can be seen in Figure P6.2b. The roots being $\delta_1, \delta_2, \delta_3, \ldots, \delta_n$. Note that $\delta_1$ lies between 0 and $\pi/2$, $\delta_2$ lies between $\pi$ and $3\pi/2$, $\delta_3$ lies between $2\pi$ and $5\pi/2$, and so on. Subtracting $T_\infty$ from Equation P6.6a and dividing by $T_0 - T_\infty$, we obtain Equation P6.6c:

$$\text{TRATIO} = \frac{T\left(\frac{x}{L},t\right) - T_\infty}{T_0 - T_\infty} = 2\sum_{n=1}^{\infty} \frac{\sin(\delta_n)\cos\left(\delta_n \frac{x}{L}\right) e^{-a\delta_n^2 t/L^2}}{\cos(\delta_n)\sin(\delta_n) + \delta_n} \quad \text{(P6.6c)}$$

A plot of TRATIO versus time, for several different values of $x/L$ should appear as shown in Figure P6.2c.

Finally, the heat transfer ratio, Qratio, from the plate to the bath in time $t$ is given by

$$\text{QRATIO} = \frac{Q(t)}{Q_0} = \frac{2hL}{k}\sum_{n=1}^{\infty} \frac{\sin\delta_n \cos\delta_n}{\delta_n^2 [\sin\delta_n \cos\delta_n + \delta_n]} \left[1 - e^{-at\delta_n^2/L^2}\right] \quad \text{(P6.6d)}$$

where:
 $Q(t)$ is the amount of heat transferred from the plate to the bath in time $t$.
 $Q_0$ is the amount of heat transferred from the plate to the bath in infinite time, which equals the change in internal energy in infinite time.

1. Write a computer program that will solve for the roots, $\delta_1, \delta_2, \ldots, \delta_{50}$ using MATLAB's fzero function. Print out the $\delta$ values in 10 rows and 5 columns. Also print out the functional values at the roots, that is, $f(\delta_n)$.

**NOTE:** Only 50 $\delta$ values were asked to be computed.

# Roots of Algebraic and Transcendental Equations

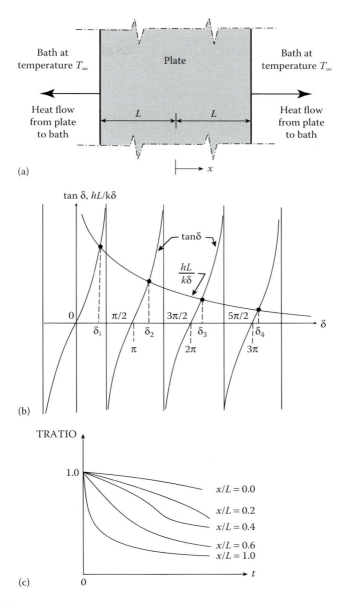

**FIGURE P6.2**
(a) Plate cooling in a bath. (b) Plot of $\tan\delta$ and $\dfrac{hL}{k\delta}$ versus $\delta$. (c) Plot of TRATIO versus $t$.

2. Solve Equation P6.6c for TRATIO for $x/L = 0.0, 0.2, 0.4, 0.6, 0.8, 1.0$ and $t = 0, 10, 20, \ldots 200$ s. Print out results in table form as shown in Table P6.4. Also use MATLAB to produce a plot similar to Figure P6.2c.

### TABLE P6.4
Temperature Ratio, TRATIO

| | | | X/L | | | |
|---|---|---|---|---|---|---|
| Time(s) | 0.0 | 0.2 | 0.4 | 0.6 | 0.8 | 1.0 |
| 0 | 1.0 | 1.0 | 1.0 | 1.0 | 1.0 | 1.0 |
| 10 | --- | --- | --- | --- | --- | --- |
| 20 | --- | --- | --- | --- | --- | --- |
| --- | --- | --- | --- | --- | --- | --- |
| 200 | --- | --- | --- | --- | --- | --- |

3. Construct a table for QRATIO versus $t$ for times 0, 10, 20, 30, ...., 200 s.
4. Use MATLAB to produce a plot of QRATIO versus $t$.

Use the following values for the parameters of the problem:

$$T_0 = 300°C, \quad T_\infty = 30°C, \quad h = 45 \text{ W/m}^2\text{-°C},$$
$$k = 10.0 \text{ W/m-°C}, \quad L = 0.03 \text{ m}, \quad a = 0.279 \times 10^{-5} \text{ m}^2/\text{s}$$

**P6.7.** In this project we consider a semiinfinite slab (such as thick layer of ice) having a uniform temperature, $T_i$, that is suddenly subjected to a change in air temperature caused by a warm front moving in over the region of interest (see Figure P6.3). The surface temperature, $T_s$, of the slab will be a function time, $t$. It will also depend on the parameters: $h$, $T_i$, $T_\infty$, $k$, and $\alpha$, where $h$ is the convective heat transfer coefficient, $T_i$ is the initial temperature of the slab and $T_\infty$ is the air temperature, $k$ and $\alpha$ are the thermal conductivity and diffusivity of the slab material respectively. The governing equation describing the surface temperature, $T_s$, as a function of time [2] is:

$$1 - e^{\frac{h^2 \alpha t}{k^2}} \left( 1 - erf\left(\frac{h\sqrt{\alpha t}}{k}\right) \right) - \frac{T_s - T_i}{T_\infty - T_i} = 0 \quad (P6.7)$$

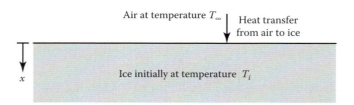

**FIGURE P6.3**
Ice slab subjected to warm front.

Given: $T_i = -20°C$, $T_\infty = 20°C$, $k = 2.22$ W/m-C, $\alpha = 12.6 \times 10^{-7}$ m$^2$/s, and $h = 100$ W/m$^2$-C. We wish to determine the time, $t$, when the surface temperature of the slab reaches the following temperatures:

$$T = [-10 \quad -5 \quad 0]°C$$

Use the search method to find an interval in which the root of Equation P6.7 lies. Then use MATLAB's `fzero` function to solve for the time, $t$, for each condition and print the results in a table with table headings. Assume that $0 \le t \le 1000$ s with a step size of 10 s.

**P6.8.** A wood circular cylinder, having a specific gravity, $S$, floats in water as shown in Figure P6.4. For a floating body, the weight of the floating body equals the weight of fluid displaced. An equation that describes the depth, $d$, of the submerged part of the floating cylinder is given by

$$f(x) = (x-1)\sqrt{2x-x^2} + \sin^{-1}(x-1) - (S-0.5)\pi = 0 \qquad (P6.8)$$

where $x = d/R$ and $S$ is the specific gravity of the wood.
 For a complete derivation of Equation P6.8, see Project P5.3 in Reference 2. Use the following parameters: $R = 0.5$ m and $0.3 \le S \le 0.5$ in steps of 0.05. Create a MATLAB program that

1. Selects a proper range for $x$ by observing Figure P6.4.
2. Uses the search method to find a small interval containing the root, use step sizes of 0.05 m.

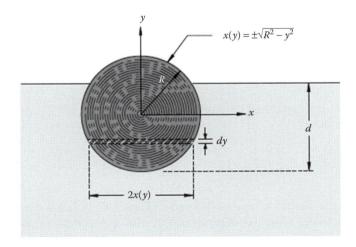

**FIGURE P6.4**
Floating wood circular cylinder.

3. Uses MATLAB's `fzero` function to obtain a better value for the root.
4. Create a table consisting of S and d (include table headings) and a plot of d versus S.

**P6.9.** The velocity of the piston described in Project 2.8 is

$$V(t) = -2\pi\omega r \sin(2\pi\omega t) - \frac{2\pi\omega r^2 \sin(2\pi\omega t)\cos(2\pi\omega t)}{\sqrt{b^2 - r^2 \sin^2(2\pi\omega t)}} \quad (P6.9)$$

Take $r = 9$ cm, $\omega = 100$ revolutions per second, $b = 14$ cm.

Create a MATLAB program that will determine the time when the velocity of the piston, described in that project, reaches 4000 cm/s during the time span $0 \le t \le 0.01$ s. Use 50 subdivisions in the $t$ domain. Print the values to the screen.

**P6.10.** In the time span $0 \le t \le 0.5$ ms, determine the number of roots and their values of the voltage, $v(t)$, of the underdamped parallel RLC circuit described in Project 2.8. The governing equation for $v(t)$ is

$$v(t) = \exp\left(-\frac{1}{2RC}t\right)\left\{A\cos\left(\sqrt{\frac{1}{LC} - \left(\frac{1}{2RC}\right)^2}\,t\right) + B\sin\left(\sqrt{\frac{1}{LC} - \left(\frac{1}{2RC}\right)^2}\,t\right)\right\} \quad (P6.10)$$

1. Use the search method to find a small interval in which each root lies.
2. In each found interval, use MATLAB's fzero function to find the root value.
3. Print out to the screen, the root number and the root value.

Assume the following parameters:

$R = 100\ \Omega$, $L = 10^{-3}$ H, $C = 10^{-6}$ F, $A = 6.0$ V, $B = -9.0$ V

$0 \le t \le 5.0 \times 10^{-4}$ s in steps of $1.0 \times 10^{-5}$ s.

**P6.11.** The current–voltage relationship of a semiconductor PN diode can be written as follows [3]:

$$V_{in} - I_S\left(e^{\frac{q}{kT}v_D} - 1\right)R_D - v_D = 0 \quad (P6.11a)$$

where $v_D$ is the diode voltage as defined in Figure P6.5, $I_S$ is a constant (with units of amperes), which is determined by the semiconductor doping concentrations and the device geometry, $q = 1.6 \times 10^{-19}$ coulomb is the unit electric charge, $k = 1.38 \times 10^{-23}$ J/K is the Boltzmann constant, and $T$ is absolute temperature (in K). We seek the value of $v_D$ that satisfies Equation P6.11a.

# Roots of Algebraic and Transcendental Equations

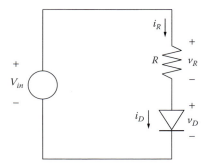

**FIGURE P6.5**
Semiconductor for PN diode.

Let

$$f(v_D) = V_{in} - I_s \left( e^{\frac{q}{kT} v_D} - 1 \right) R - v_D = 0 \tag{P6.11b}$$

1. Create a MATLAB function for $f(v_D)$ and plot for the interval $0 \leq v_D \leq 0.8$ V for 10 mV steps (80 subdivisions on the $v_D$ domain).
2. Use the search method to obtain a small interval within which the root of Equation P6.11b lies.
3. Use MATLAB's fzero function to obtain a more accurate value for the root. Use the following parameters:

$$T = 300 \text{ K}, I_s = 10^{-14} \text{ A}, V_{in} = 5 \text{ V}, R = 1000 \, \Omega.$$

4. Print the root value to the screen.

**P6.12.** We wish to determine the DC transfer characteristic for the diode circuit of Figure P6.5. We will consider $V_{in}$ as a parameter, where $5 \leq V_{in} \leq 12$ in steps of 1 V. We wish to find the value of $v_D$ for all values of $V_{in}$.

Write a MATLAB program that will find the roots of $f(v_D) = 0$, where

$$f(v_D) = V_{in} - I_s \left( e^{\frac{qv_D}{kT}} - 1 \right) R - v_D = 0 \tag{P6.12}$$

Your program should

1. Take $0.2 \leq v_D \leq 0.8$ V with 60 subdivisions on the $v_D$ domain.
2. Use the search method to find a small interval in which the root of $f(v_D) = 0$ lies.
3. Use the MATLAB's fzero function to obtain a more accurate value for the root.

4. Construct a table consisting of all values of $V_{in}$ and the corresponding roots of $f(v_D) = 0$.
5. If you did Project 6.11, confirm that the root value obtained in this project when $V_{in} = 5$ V is the same as that obtained in Project 6.11.

## References

1. Moran, M. J. and Shapiro, H. N., *Fundamentals of Thermodynamics*, John Wiley & Sons, Hoboken, NJ, 2004.
2. Bober, W., *Introduction to Numerical and Analytical Methods with MATLAB for Engineers and Scientists*, CRC Press, Boca Raton, FL, 2014.
3. Bober, W. and Stevens, A., *Numerical and Analytical Methods with MATLAB for Electrical Engineers*, CRC Press, Boca Raton, FL, 2013.

# 7

# System of Algebraic, Linear Equations

## 7.1 Introduction

In engineering we are frequently confronted with dealing with a problem involving a set of algebraic, linear equations. In this chapter we discuss the use of MATLAB®'s inv and Gauss-Elimination functions for solving a system of algebraic, linear equations.

Before the use of computers, the method of determinants was used to obtain a solution to a system of algebraic, linear equations. Computationally, it is only practical for a system involving just a few equations [1]. Since it is much easier to solve such a system by MATLAB's inv function or MATLAB's Gauss-Elimination function, we will skip the method of determinants.

## 7.2 System of Algebraic, Linear Equations

Given the set of equations

$$a_{1,1}x_1 + a_{1,2}x_2 + a_{1,3}x_3 + \cdots + a_{1,n}x_n = c_1$$

$$a_{2,1}x_1 + a_{2,2}x_2 + a_{2,3}x_3 + \cdots + a_{2,n}x_n = c_2$$

$$\vdots$$

$$a_{n,1}x_1 + a_{n,2}x_2 + a_{n,3}x_3 + \cdots + a_{n,n}x_n = c_n$$

(7.1)

where the $a$'s and the $c$'s are known and the $x$'s are the unknowns. In matrix algebra we can write the set of Equations 7.1 as follows:

$$\mathbf{AX} = \mathbf{C}$$

where:

$$X = \begin{bmatrix} x_1 \\ x_2 \\ \vdots \\ x_n \end{bmatrix}, \quad A = \begin{bmatrix} a_{1,1} & a_{1,2} & \cdots & a_{1,n} \\ a_{2,1} & a_{2,2} & \cdots & a_{2,n} \\ \vdots & \vdots & \ddots & \vdots \\ a_{n,1} & a_{n,2} & \cdots & a_{n,n} \end{bmatrix}, \quad C = \begin{bmatrix} c_1 \\ c_2 \\ \vdots \\ c_n \end{bmatrix} \quad (7.2)$$

The unknown vector **X** has $n$ rows and 1 column. Similarly, the known vector **C** has $n$ rows and 1 column. The known coefficient matrix **A** has $n$ rows and $n$ columns.

NOTE: In matrix algebra, the number of columns in **A** must equal the number of rows in **X**.

### 7.2.1 MATLAB's `inv` Function

The solution of the set of Equations 7.1 can be obtained by the use of MATLAB's inv function as follows:

$$X = \text{inv}(A) * C \quad (7.3)$$

where:
  X is the solution to the set of Equations 7.1
  A is the coefficient matrix shown in Equation 7.2
  C is the vector shown in Equation 7.2

The method of solving a system of linear equations by use of MATLAB's inv function is more computationally complicated than a method called Gauss elimination. MATLAB's method for solving the system of Equations 7.1 by the Gauss-Elimination method is shown below.

### 7.2.2 Gauss-Elimination Method

To solve the system of Equations 7.1 by MATLAB's Gauss-Elimination method use

$$X = A \backslash C \quad (7.4)$$

where X, A, and C have the same meaning as in Equation 7.3.
Note the use of MATLAB's backslash operator to solve for X by Gauss Elimination.

You can obtain the size of matrix **A** by the command **size(A)**. This command is useful when you run a script and you get an error message like "Index exceeds matrix dimensions." Entering the **size()** command in the script will help you determine the problem.

# System of Algebraic, Linear Equations 155

**Example 7.1**

The following example solves the third-order system of linear equations shown below, and writes the results to the screen:

$$3x_1 + 2x_2 - x_3 = 10$$
$$-x_1 + 3x_2 + 2x_3 = 5$$
$$x_1 - x_2 - x_3 = -1$$

```
% Example_7_1.m
% This program solves a third order linear system of equations by
% MATLAB's inv function and by MATLAB's Gauss elimination method.
clc; clear;
A=[3 2 -1; -1 3 2; 1 -1 -1];
C=[10 5 -1]'; % Transposing a row vector to a column vector.
% check solution:
X1=inv(A)*C % X1 is the solution using MATLAB's inv function.
X2=A\C % X2 is the solution using MATLAB's Gauss elimination method.
% Does X2=X1?
% Use the size() command to determine the number of rows and columns.
% Print size of A.
[A_rows A_cols] = size(A);
% Print matrix A.
A
% Print vector C.
C
% Print A*X1, does it give C2=C? \n');
C2=A*X1
end
```
-------------------------------------------------------------------------------

**Program Results:**

X1 =

    -2.0000
     5.0000
    -6.0000

X2 =

    -2.0000
     5.0000
    -6.0000

A =

     3    2   -1
    -1    3    2
     1   -1   -1

```
C =

    10
     5
    -1

C2 =

    10.0000
     5.0000
    -1.0000
>>
```

We see that the use of MATLAB's Gauss-Elimination method produces the same answer as the use of MATLAB's `inv` function.

**Exercises**

**E7.1.** Solve the following set of linear equations by MATLAB's `inv` function:

a. $2x_1 - x_2 = 12$
   $4x_1 + 3x_2 = -8$

b. $2x_1 + 3x_2 - x_3 = 20$
   $4x_1 - x_2 + 3x_3 = -14$
   $x_1 + 5x_2 + x_3 = 21$

c. $4x_1 + 8x_2 + x_3 = 8$
   $-2x_1 - 3x_2 + 2x_3 = 14$
   $x_1 + 3x_2 + 4x_3 = 30$

## 7.3 Treatment of Large Systems of Algebraic, Linear Equations

When there is a large system of algebraic, linear equations, it may not be obvious how to determine the appropriate matrices. There is a systematic approach that can be used. This is demonstrated in the following example:

# System of Algebraic, Linear Equations

## Example 7.2

Suppose we have the following system of equations (this system of equations was determined by the analysis of a truss consisting of 13 structural members subjected to external forces). In the following set of equations, $F_1, F_2, F_3, \ldots F_{13}$ represent the unknown internal forces in the structural members and the right-hand side of the equations represent external forces, $P_1, P_2, P_3, \ldots P_{13}$ applied to the truss members at the joints. (For more details on the process see Section 4.4 in Reference 2). Since there are 13 unknown forces, there will be 13 equations and each equation will have 13 $a$'s, most of which will be zero. The equations need to be numbered as shown below. In the following set of equations, the coefficient matrix **A** is be made up of elements $a_{i,j}$ where the first index is the equation number and the second index is the same number, $j$, of the index of unknown force, $F_j$.

$0.6\ F_1 + F_2 = 0$     (1) We see that only $a_{1,1}$ and $a_{1,2}$ are non-zero.

$0.8\ F_1 = -62.5$     (2) We see that only $a_{2,1}$ is non-zero.

$-0.6\ F_1 + F_5 = 0$     (3) We see that only $a_{3,1}$ and $a_{3,5}$ are non-zero.

$-0.8\ F_1 - F_3 = 0$     (4) We see that only $a_{4,1}$ and $a_{4,3}$ are non-zero.

$-F_2 + F_6 + 0.6\ F_4 = 0$     (5) We see that only $a_{5,2}, a_{5,6}$ and $a_{5,4}$ are non-zero.

$F_3 + 0.8\ F_4 = 50$     (6) We see that only $a_{6,3}$ and $a_{6,4}$ are non-zero.

$-0.6\ F_4 - F_5 + F_8 + 0.6\ F_{10} = 0$     (7) We see that only $a_{7,4}, a_{7,5}, a_{7,8}$ and $a_{7,10}$ are non-zero.

$-0.8\ F_4 - F_7 - 0.8\ F_{10} = 0$     (8) We see that only $a_{8,4}, a_{8,7}$ and $a_{8,10}$ are non-zero.

$-F_6 + F_9 = 0$     (9) We see that only $a_{9,6}$ and $a_{9,9}$ are non-zero.

$F_7 = 30$     (10) We see that only $a_{10,7}$ are non-zero.

$-F_9 - 0.6\ F_{10} + F_{11} = 0$     (11) We see that only $a_{11,9}, a_{11,10}$ and $a_{11,11}$ are non-zero.

$0.8\ F_{10} + F_{12} = 40$     (12) We see that only $a_{12,10}$ and $a_{12,12}$ are non-zero.

$-F_{12} - 0.8\ F_{13} = 0$     (13) We see that only $a_{13,12}$ and $a_{13,13}$ are non-zero.

Since the coefficient matrix is sparse, it is best to initially set all $a_{i,j} = 0$, and then overwrite the non-zero $a_{i,j}$ terms as specified in the set of equations. In matrix algebra, the set of equations is of the form AF = P. The program follows:

```
% Example_7_2.m
% This program solves a system of algebraic, linear equations.
% The system of equations stems from a truss problem in statics.
% The A matrix elements are initially set to zero. The non-zero
% A elements then override the initial values. The first index in the A
% matrix elements represent the equation number. The second index in
% the A matrix element correspond to the index of the force associated with
% that matrix element.
```

```matlab
% There are 13 equations for 13 unknown internal forces, F(1)-F(13).
clear; clc;
ie=13; je=13;
a=zeros(13);
p=zeros(13,1);
fo=fopen('output.txt','w');
fprintf(fo,'Example 7.2 \n');
fprintf(fo,'Program solves for the internal forces of a truss');
fprintf(fo,'Forces in kN \n');
% Overwrite the non-zero elements of matrix a and matrix p.
a(1,1)=0.6; a(1,2)= 1; p(1)=0; % From Equation 1.
a(2,1)=0.8; p(2)=-62.5;        % From Equation 2, etc.
a(3,1)=-0.6; a(3,5)=1; p(3)=0;
a(4,1)=-0.8; a(4,3)=-1; p(4)=0;
a(5,2)=-1.0; a(5,6)=1; a(5,4)=0.6; p(5)=0;
a(6,3)=1.0; a(6,4)=0.8; p(6)=50.0;
a(7,4)=-0.6; a(7,5)=-1.0; a(7,8)=1.0; a(7,10)=0.6; p(7)=0;
a(8,4)=-0.8; a(8,10)=-0.8; a(8,7)=-1; P(8)=0;
a(9,6)=-1; a(9,9)=1.0; p(9)=0;
a(10,7)=1.0; p(10)=30;
a(11,9)=-1.0; a(11,10)=-0.6; a(11,11)=1.0; p(11)=0;
a(12,12)=1.0; a(12,10)=0.8; p(12)=40;
a(13,12)=-1.0; a(13,13)=-0.8; p(13)=0;
fprintf(fo,' A matrix \n\n');
jindex=1:je;
fprintf(fo,' ');
for i=1:ie
    fprintf(fo,'%5i',jindex(i));
end
fprintf(fo,'\n');
fprintf(fo,'-------------------------------------------');
fprintf(fo,'----------------\n');
for i=1:ie
    fprintf(fo, '%4i' ,i);
    for j=1:je
        fprintf(fo,'%5.1f',a(i,j));
    end
    fprintf(fo,'\n');
end
F=a\p;
fprintf(fo,'\n\n');
fprintf(fo,'Internal forces, F(1)-F(13)& external forces p(i) \n\n');
fprintf(fo,'Member No.    F(kN)     Equation No    p(kN)   \n');
fprintf(fo,'===================================================\n');
for i=1:ie
    fprintf(fo,'    %3.0f      %9.2f \t    %3.0f \t\t     %5.1f \n',...
        i,F(i),i,p(i));
end
```

# System of Algebraic, Linear Equations

**Program Results:**

```
Example 7.2
Program solves for the internal forces of a truss. Forces in kN.
A matrix
        1     2     3     4     5     6     7     8     9    10    11    12    13
   ---------------------------------------------------------------------------------
 1    0.6   1.0   0.0   0.0   0.0   0.0   0.0   0.0   0.0   0.0   0.0   0.0   0.0
 2    0.8   0.0   0.0   0.0   0.0   0.0   0.0   0.0   0.0   0.0   0.0   0.0   0.0
 3   -0.6   0.0   0.0   0.0   1.0   0.0   0.0   0.0   0.0   0.0   0.0   0.0   0.0
 4   -0.8   0.0  -1.0   0.0   0.0   0.0   0.0   0.0   0.0   0.0   0.0   0.0   0.0
 5    0.0  -1.0   0.0   0.6   0.0   1.0   0.0   0.0   0.0   0.0   0.0   0.0   0.0
 6    0.0   0.0   1.0   0.8   0.0   0.0   0.0   0.0   0.0   0.0   0.0   0.0   0.0
 7    0.0   0.0   0.0  -0.6  -1.0   0.0   0.0   1.0   0.0   0.6   0.0   0.0   0.0
 8    0.0   0.0   0.0  -0.8   0.0   0.0  -1.0   0.0   0.0  -0.8   0.0   0.0   0.0
 9    0.0   0.0   0.0   0.0   0.0  -1.0   0.0   0.0   1.0   0.0   0.0   0.0   0.0
10    0.0   0.0   0.0   0.0   0.0   0.0   1.0   0.0   0.0   0.0   0.0   0.0   0.0
11    0.0   0.0   0.0   0.0   0.0   0.0   0.0   0.0  -1.0  -0.6   1.0   0.0   0.0
12    0.0   0.0   0.0   0.0   0.0   0.0   0.0   0.0   0.0   0.8   0.0   1.0   0.0
13    0.0   0.0   0.0   0.0   0.0   0.0   0.0  -1.0   0.0   0.0   0.0  -1.0   0.6

Internal forces, F(1)-F(13) & external forces p(i)

Member No.      F(kN)       Equation No    p(kN)
===================================================
    1          -78.13            1          0.0
    2           46.88            2        -62.5
    3           62.50            3          0.0
    4          -15.63            4          0.0
    5          -46.88            5          0.0
    6           56.25            6         50.0
    7           30.00            7          0.0
    8          -43.13            8          0.0
    9           56.25            9          0.0
   10          -21.88           10         30.0
   11           43.13           11          0.0
   12           57.50           12         40.0
   13          -71.88           13          0.0
```

## 7.4 A Resistive Circuit Problem

Another example involving a system of linear equations can be found in problems involving resistive circuits (see Figure 7.1). For more information on the subject see Section 4.5 in Reference 2. The goal is to solve for the node voltages $v_1$, $v_2$, and $v_3$ as functions of the input voltages $V_1$ and $V_2$ and the input current $I_1$. The equations involve the conductances, $G_1, G_2, ..., G_5$, where $G_i = 1/R_i$ and $R_i$ are the resistant members in the circuit.

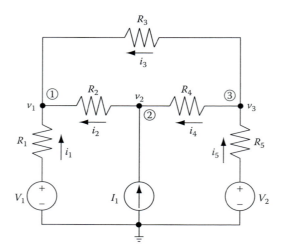

**FIGURE 7.1**
A resistive circuit for Example 7.3.

The governing equations for this example are

$$(G_1 + G_2 + G_3)v_1 - G_2 v_2 - G_3 v_3 = V_1 G_1 \tag{7.5}$$

$$-G_2 v_1 + (G_2 + G_4)v_2 - G_4 v_3 = I_1 \tag{7.6}$$

$$-G_3 v_1 - G_4 v_2 + (G_3 + G_4 + G_5)v_3 = V_2 G_5 \tag{7.7}$$

The right-hand side of Equations 7.5 through 7.7 will be represented as vector C.

This system of equations may be solved by the use of MATLAB's inv function or by MATLAB's Gauss-Elimination method as shown in the following Example 7.3.

The following script solves for the node voltages, $v_1$, $v_2$, and $v_3$ for the following circuit values:

$R_1 = 2200 \ \Omega, \ R_2 = 10 \ k\Omega, \ R_3 = 6900 \ \Omega, \ R_4 = 9100 \ \Omega, \ R_5 = 3300 \ \Omega$

$V_1 = 12 \ V, \ V_2 = 3.3 \ V, \ I_1 = 2 \ mA$

```
% Example_7_3.m
% Resistive Circuit Problem
% This program solves for the internal node voltages for the circuit
% shown in Figure 7.1.
% The conductances G are in units of Siemens.
% The node voltage V are in units of volts.
```

## System of Algebraic, Linear Equations

```
% The currents I are in units of amps.
clear; clc;
A=zeros(3);
C=zeros(3,1);
V=zeros(3,1);
g1 = 1/2200; g2 = 1/10000; g3 = 1/6900; g4 = 1/9100; g5 = 1/3300;
V1 = 12; V2 = 3.3; I1 = .002;
% Overwrite the non-zero elements of matrix A and vector C.
A(1,1)=g1+g2+g3; A(1,2)=-g2; A(1,3)=-g3; C(1)=V1*g1;
A(2,1)=-g2; A(2,2)=g2+g4; A(2,3)=-g4; C(2)=I1;
A(3,1)=-g3; A(3,2)=-g4; A(3,3)=g3+g4+g5; C(3)=V2*g5;
v = A \ C ;
% print the results
fprintf('V1=%5.1f V    V2=%5.1f V    I1=%5.1e A \n',V1,V2,I1);
fprintf('g1=%8.5f S    g2=%8.5f S    g3=%8.5f S \n',g1,g2,g3);
fprintf('g4=%8.5f S    g5=%8.5f S\n',g4,g5);
fprintf('\n\n');
fprintf('Node #       v (volts)    \n');
fprintf('--------------------------\n');
for n=1:length(C)
    fprintf('   %3i       %9.1f \n', n,v(n));
end
```
---
**Program Results:**
```
V1= 12.0 V      V2= 3.3 V        I1= 2.0e-03 A
g1= 0.00045 S   g2= 0.00010 S    g3= 0.00014 S
g4= 0.00011 S   g5= 0.00030 S

Node #       v (volts)
--------------------------
   1            12.6
   2            20.3
   3             9.0
>>
```
---

## 7.5 Gauss Elimination

As previously discussed, the Gauss-Elimination method in solving a algebraic, linear set of equations is computationally more efficient than the use of MATLAB's inv function. In the Gauss-Elimination method, the original system is reduced to an **equivalent triangular set** that can readily be **solved by back substitution** (for a complete description of the method see Section 4.6 in Reference 2). The reduced **equivalent** set would appear like the following set of equations:

$$\begin{aligned}
\tilde{a}_{11}x_1 + \tilde{a}_{12}x_2 + \tilde{a}_{13}x_3 + \cdots + \tilde{a}_{1n}x_n &= \tilde{c}_1 \\
\tilde{a}_{22}x_2 + \tilde{a}_{23}x_3 + \cdots + \tilde{a}_{2n}x_n &= \tilde{c}_2 \\
\tilde{a}_{33}x_3 + \cdots + \tilde{a}_{3n}x_n &= \tilde{c}_3 \\
\ddots \quad \vdots &= \vdots \\
\tilde{a}_{n-1,n-1}x_{n-1} + \tilde{a}_{n-1,n}x_n &= \tilde{c}_{n-1} \\
\tilde{a}_{n,n}x_n &= \tilde{c}_n
\end{aligned} \qquad (7.8)$$

where the tilde (~) variables are a new set of coefficients (to be determined), and where the new coefficient matrix $\tilde{\mathbf{A}}$ is *diagonal* (i.e., all of the coefficients left of the main diagonal are zero). Then by back substitution,

$$x_n = \tilde{c}_n / \tilde{a}_{n,n}$$

$$x_{n-1} = \frac{1}{\tilde{a}_{n-1,n-1}} (\tilde{c}_{n-1} - \tilde{a}_{n-1,n}x_n)$$

$$\vdots$$

etc.

## 7.6 Number of Solutions

Suppose a Gauss-Elimination program is carried out and the following results are obtained:

$$\begin{aligned}
a_{11}x_1 + a_{12}x_2 + a_{13}x_3 + \cdots + a_{1n}x_n &= c_1 \\
a_{22}x_2 + a_{23}x_3 + \cdots + a_{2n}x_n &= c_2 \\
a_{33}x_3 + \cdots + a_{3n}x_n &= c_3 \\
\ddots \quad \vdots &= \vdots \\
a_{rr}x_r + \cdots + a_{rn}x_n &= c_r \\
0 &= c_{r+1} \\
0 &= c_{r+2} \\
\vdots \\
0 &= c_n
\end{aligned} \qquad (7.9)$$

# System of Algebraic, Linear Equations

where $r < n$ and $a_{11}, a_{22}, \ldots, a_{rr}$ are not zero. There are two possible cases:

1. No solution exists if any one of the $c_{r+1}$ through $c_n$ is not zero.
2. Infinitely many solutions exits if $c_{r+1}$ through $c_n$ are all zero.

NOTE: If you attempt to solve a system of algebraic, linear equations using MATLAB's Gauss-Elimination method, and MATLAB arrives at a set of equations as shown in Equation 7.9, MATLAB will give you a warning as follows:

"Warning: Matrix is close to singular or badly scaled. Results may be inaccurate."

If, in the above set, $r = n$ and $a_{11}, a_{22}, \ldots, a_{nn}$ are not zero, then the system would appear as follows:

$$\begin{aligned}
a_{11}x_1 + a_{12}x_2 + a_{13}x_3 + \cdots + a_{1n}x_n &= c_1 \\
a_{22}x_2 + a_{23}x_3 + \cdots + a_{2n}x_n &= c_2 \\
a_{33}x_3 + \cdots + a_{3n}x_n &= c_3 \\
\ddots \quad \vdots &= \vdots \\
a_{nn}x_n &= c_n
\end{aligned} \quad (7.10)$$

For this case there is only one solution.

**Exercise**

**E7.2.** Use MATLAB's `inv` function to solve the following set of linear equations. Note the warning that MATLAB gives with its solution.

$$2x_1 + 3x_2 + x_3 - x_4 = 1$$
$$5x_1 - 2x_2 + 5x_3 - 4x_4 = 5$$
$$x_1 - 2x_2 + 3x_3 - 3x_4 = 3$$
$$3x_1 + 8x_2 - x_3 + x_4 = -1$$

---

## REVIEW 7.1

1. Given a set of algebraic, linear equations in the form $\mathbf{AX} = \mathbf{C}$, where $\mathbf{A}$ is the coefficient matrix and $\mathbf{X}$ and $\mathbf{C}$ are column vectors, what are the two ways for solving for $\mathbf{X}$ in MATLAB?
2. Given a large system of algebraic, linear equations of the form $\mathbf{AX} = \mathbf{C}$, describe the recommended approach to solving the system of linear equations.

## Projects

**P7.1.** The following set of linear equations came from a problem in Statics. Use the method described in Example 7.2 to solve the following set of linear equations. The $F_i$ value represents the internal force in structural member $i$ in kN.

Take $A_x = -9$ kN, $A_y = 7$ kN, $\cos \vartheta = 0.6$, $\sin \vartheta = 0.8$

$$\cos \vartheta F_1 + F_2 = -A_x \quad (1)$$
$$\sin \vartheta F_1 = -A_y \quad (2)$$
$$-\cos \vartheta F_1 + \cos \vartheta F_5 + F_6 = 0 \quad (3)$$
$$-\sin \vartheta F_1 - F_3 - \sin \vartheta F_5 = 0 \quad (4)$$
$$-F_2 + F_4 = 0 \quad (5)$$
$$F_3 = 2 \quad (6)$$
$$-F_6 + \cos \vartheta F_9 + F_{10} = 0 \quad (7)$$
$$-F_7 - \sin \vartheta F_9 = 0 \quad (8)$$
$$-F_4 - \cos \vartheta F_5 + F_8 = 0 \quad (9)$$
$$\sin \vartheta F_5 + F_7 = 4 \quad (10)$$
$$-F_{10} + F_{14} = 0 \quad (11)$$
$$F_{11} = 0 \quad (12)$$
$$-F_8 - \cos \vartheta F_9 + F_{12} + \cos \vartheta F_{13} = 0 \quad (13)$$
$$\sin \vartheta F_9 + F_{11} + \sin \vartheta F_{13} = 6 \quad (14)$$
$$-\cos \vartheta F_{13} - F_{14} + F_{18} = 0 \quad (15)$$
$$-\sin \vartheta F_{13} - F_{15} = 0 \quad (16)$$
$$-F_{12} + F_{16} + \cos \vartheta F_{17} = 0 \quad (17)$$
$$F_{15} + \sin \vartheta F_{17} = 4 \quad (18)$$
$$-\cos \vartheta F_{17} - F_{18} + \cos \vartheta F_{21} = -9 \quad (19)$$
$$-\sin \vartheta F_{17} - F_{19} - \sin \vartheta F_{21} = 0 \quad (20)$$
$$-F_{16} + F_{20} = 0 \quad (21)$$

# System of Algebraic, Linear Equations

**P7.2.** The following set of linear equations came from a problem in Statics. Use the method described in Example 7.2 to solve the following set of linear equations. The $F_i$ value represents the internal force in structural member $i$ in kN.

$$0.8F_2 = 30 \quad (1)$$
$$F_1 + 0.6F_2 = -30 \quad (2)$$
$$F_3 + 0.8F_5 = 0 \quad (3)$$
$$-F_1 + F_4 + 0.6F_5 = 0 \quad (4)$$
$$-0.8F_2 - F_3 = 5 \quad (5)$$
$$-0.6F_2 + F_6 = 0 \quad (6)$$
$$F_7 + 0.8F_9 - F_{11} = 0 \quad (7)$$
$$-F_4 + 0.6F_9 + F_{11} = 0 \quad (8)$$
$$-0.8F_5 - F_7 = 5 \quad (9)$$
$$-0.6F_5 - F_6 + F_8 = 0 \quad (10)$$
$$0.8F_9 + F_{10} = -5 \quad (11)$$
$$-F_8 - 0.6F_9 = 0 \quad (12)$$
$$-F_{10} + 0.8F_{13} + F_{14} = 0 \quad (13)$$
$$-F_{11} - 0.6F_{13} = 0 \quad (14)$$
$$F_{12} + 0.8F_{13} - F_{16} = 0 \quad (15)$$
$$0.6F_{13} + F_{15} = 15 \quad (16)$$
$$F_{16} + 0.8F_{17} - F_{18} = 0 \quad (17)$$
$$0.6F_{17} = 15 \quad (18)$$

**P7.3.** Figure P7.1 shows a resistive circuit known as a *ladder network*.

Using Ohm's law and Kirchoff's current law, we can determine a set of linear equations for the voltages $v_1, v_2, v_3$, and $v_4$ (For a complete derivation of the set of equations, see Project P4.5 in Reference 2). The equations are written in terms of the conductances, $G_{nm}$, instead of the resistances, $R_{nm}$, where $G_{nm} = 1/R_{nm}$.

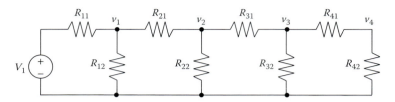

**FIGURE P7.1**
Fourth-order ladder network.

The governing equations for this system are

$$-(G_{11}+G_{21}+G_{12})v_1 + G_{21}v_2 = -G_{11}V_{ref} \quad (1)$$

$$G_{21}v_1 - (G_{21}+G_{31}+G_{22})v_2 + G_{31}v_3 = 0 \quad (2)$$

$$G_{31}v_2 - (G_{31}+G_{41}+G_{32})v_3 + G_{41}v_4 = 0 \quad (3)$$

$$G_{41}v_3 - (G_{41}+G_{42})v_4 = 0 \quad (4)$$

Create a MATLAB program to solve for all circuit voltages. Take $V_1 = 5V$ and the following resistor values:

$$R_{11} = 2200 \, \Omega, \; R_{12} = 2200 \, \Omega$$

$$R_{21} = 1200 \, \Omega, \; R_{22} = 6800 \, \Omega$$

$$R_{31} = 3900 \, \Omega, \; R_{32} = 2200 \, \Omega$$

$$R_{41} = 3300 \, \Omega, \; R_{42} = 5700 \, \Omega$$

**P7.4.** Suppose a manufacturer wishes to purchase a piece of equipment that costs $40,000. He plans to borrow the money from a bank and pay off the loan in 10 years in 120 equal payments. The annual interest rate is 6%. Each monthly payment, M, consists of two parts: one part goes toward paying off the principal, P, and the other part is the interest charged based on the unpaid balance of the loan. He wishes to determine what his monthly payment will be. This problem can be solved by a system of linear equations. Let $x_j$ is the amount in the *jth* payment that goes toward paying off the principal. Then the equation describing the *jth* payment is

$$jth\,payment = M = x_j + \left(P - \sum_{n=1}^{n=j-1} x_n\right)I \quad (P7.4a)$$

where:
M is the monthly payment
P is the amount borrowed
I is the monthly interest rate = annual interest rate/12

The total number of unknowns is 121 (120x values and M).

Applying Equation P7.4a to each month gives 120 equations. One additional equation is

$$P = \sum_{n=1}^{n=120} x_n \qquad \text{(P7.4b)}$$

Develop a MATLAB program that will
1. Ask the user to enter from the keyboard the amount of the loan (P), the annual interest rate, $I$, and the time period, $Y$, in years.
2. Set up the system of linear equations, using $A_{n,m}$ as the coefficient matrix of the system of linear equations. The $n$ represents the equation number and $m$ represents the coefficient of $x_m$ in that equation. Set $x_{121} = M$.
3. Solve the system of linear equations in MATLAB.
4. Print out a table consisting of four columns. The first column should be the month number, the second column the monthly payment, the third column the amount of the monthly payment that goes toward paying off the principal, and the fourth column the interest payment for that month.

## References

1. Kreyszig, E., *Advanced Engineering Mathematics*, 8th ed., Wiley, New York, 1999.
2. Bober, W., *Introduction to Numerical and Analytical Methods with MATLAB for Engineers and Scientists*, CRC Press, Boca Raton, FL, 2014.

# 8
# Curve Fitting

## 8.1 Introduction

There are many occasions in engineering that require an experiment to determine the behavior of a particular phenomenon. The experiment may produce a set of data points that represents a relationship between the variables involved in the phenomenon. We may then wish to express this relationship analytically for further analysis. A mathematical expression that describes the data is called an *approximating function*. There are two approaches to determining an approximating function:

1. The approximating function graphs as a smooth curve. The approximating curve will generally not pass through all the data points, but we seek to minimize the resulting error in order to get the best fit. A plot of the data on linear, semilog or log-log graphic paper can often suggest an appropriate form for the approximating function.
2. The approximating function passes through all data points (as described in Section 8.4). However, if there is some scatter in the data points, this approximating function may not be satisfactory.

## 8.2 MATLAB's Curve-Fitting Functions

MATLAB® calls curve fitting with a polynomial by the name *Polynomial Regression*. The function polyfit(x, y, m) returns a vector of (m + 1) coefficients, $a_i$, that represent the best-fit polynomial of degree $m$ for the $(x_i, y_i)$ set of $n$ data points. The coefficient order corresponds to decreasing powers of $x$, that is,

$$y_c = a_1 x^m + a_2 x^{m-1} + a_3 x^{m-2} + \ldots a_m x + a_{m+1} \qquad (8.1)$$

To obtain $y_c$ at the data points $(x_1, x_2, x_2..., x_n)$ use the MATLAB function polyval(a, x), where $x = [x_1\ x_2\ ....\ x_n]$. MATLAB's polyval(a, x) function returns a vector of length $n$ giving $y_{c,i}$ where

$$y_{c,i} = a_1 x_i^m + a_2 x_i^{m-1} + a_3 x_i^{m-2} + ... a_m x_i + a_{m+1} \qquad (8.2)$$

MATLAB measures the precision of the fit with a function named mse, which is defined as follows:

$$\text{mse} = \frac{1}{n}\sum_{i=1}^{n}(y_i - y_{c,i})^2 \qquad (8.3)$$

where $y_i$ are the data point $y$ values and $y_{c,i}$ are the approximating curve $y$ values at the data points $x_i$, and $n$ is the number of data points.

You may also use polyval(a, x), where x is any set of x values, preferably between $x_1$ and $x_n$.

**Example 8.1**

```
% Example_8_1.m
% This program determines the best fit polynomial approximating
% functions of orders 2 thru 5 for the data set listed below.
% MATLAB's polyfit and polyval functions are used in the program.
% The sprintf command is used in this program to write formatted data
% in the plot title. The sprintf command is the same as the
% fprintf command except that it returns the data in a MATLAB
% string rather than writing to the screen or to a file.
clear; clc;
% Enter data.
x=-10:2:10;
y=[-980 -620 -70 80 100 90 0 -80 -90 10 220];
mse=zeros(4);
% Determine best fit 2-5 degree polynomials to fit the data.
for m=2:5
    fprintf('m= %i \n',m);
    coef=zeros(m+1);
    coef=polyfit(x,y,m);
    % Approximating function at x
    yc=polyval(coef,x);
    % yc is a vector since x is a vector.
    mse(m)=sum((y-yc).^2)/length(x);
    % y-yc=[y(1)-yc(1) y(2)-yc(2) ..... y(n)-yc(n)].
    fprintf('   x            y           yc \n');
    fprintf('-----------------------------\n');
    for i=1:length(x)
        fprintf('%5.1f    %5.1f    %8.2f \n',x(i),y(i),yc(i));
```

## Curve Fitting

```
        end
        fprintf('\n\n');
        x2=-10:0.5:10;
        % Approximating function at x2
        yc2=polyval(coef,x2);
        subplot(2,2,m-1),plot(x2,yc2,x,y,'o'), xlabel('x'), ylabel('y'),
        grid, axis([ -10 10 -1500 500]),
        legend('approx curve','data points');
        title(sprintf('Degree %d polynomial fit',m));
end
fprintf(' m      mse    \n')
fprintf('-----------------\n');
for m=2:5
    fprintf('%d     %8.1f \n',m,mse(m))
end
```
------------------------------------------------------------------------------


### Program Results:
Output for m = 5 is only displayed here.
```
m = 5
    x          y          yc
-------------------------------
 -10.0      -980.0      -999.09
  -8.0      -620.0      -545.31
  -6.0       -70.0      -156.76
  -4.0        80.0        78.39
  -2.0       100.0       148.18
   0.0        90.0        93.80
   2.0         0.0       -13.50
   4.0       -80.0       -95.45
   6.0       -90.0       -89.91
   8.0        10.0        26.15
  10.0       220.0       213.50


    m       mse
   -----------------
    2     32842.4
    3      2660.0
    4      2342.1
    5      1502.9
>>
```

As expected, the mse decreases as the order of the fitted polynomial is increased.
See Figure 8.1.
------------------------------------------------------------------------------

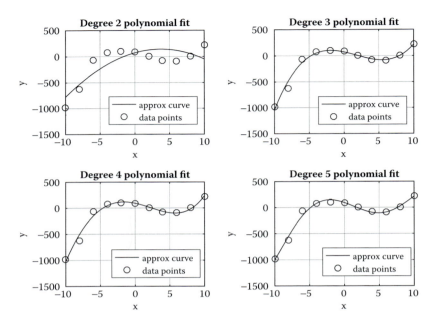

**FIGURE 8.1**
Approximating curves with data points.

**Example 8.2**

This example involves the sampling with respect to time of an audio signal that is converted to a voltage by a microphone and an amplifier. The data are in volts (V) versus time in microseconds (μs).

```
% Example_8_2.m
% This program determines the best fit polynomial approximating
% functions of orders 2 thru 5 for the data set listed below.
% MATLAB's polyfit and polyval functions are used in the program.
% The data involves the sampling in time, t, of an audio signal
% converted to a voltage, V, by a microphone and an amplifier.
% V is in volts and t is in microseconds.
clear; clc;
% Define original data points for V(t)
t = [ 0 4 8 12 16 20 24 28 32 36 40 44 48 52 56 ];
V = [.7 .9 .9 .7 .3 0 -.3 -.7 -.7 -.3 0 .3 .7 .7 .3 ];
% Determine best fit 2-5 degree polynomials to fit the data.
for m=2:5
    fprintf('m= %i \n',m);
    coef=zeros(m+1);
    coef=polyfit(t,V,m);
    % Approximating function at t
    Vc=polyval(coef,t);
    mse(m)=sum((V-Vc).^2)/length(t);
```

# Curve Fitting

```
        fprintf('     t                V              Vc     \n');
        fprintf(' (micro-sec)       (volt)          (volt) \n');
        fprintf('-------------------------------------------\n');
        for i=1:length(t)
            fprintf('  %5.1f        %8.4f        %8.4f \n',...
                t(i),V(i),Vc(i));
        end
        fprintf('\n\n');
        t2 = 0:60;
        % Approximating function at t2
        Vc2=polyval(coef,t2);
        subplot(2,2,m-1),plot(t2,Vc2,t,V,'o'), xlabel('t(\mus)'),
        ylabel('V(V)'), grid, legend('approx curve','data points'),
        title(sprintf('Degree %d polynomial fit',m));
end
fprintf('m        mse   \n')
fprintf('--------------------\n');
for m=2:5
    fprintf('%d         %8.5f \n',m,mse(m))
end
```
---

**Program Results:**

Only results for m = 5 are displayed.
```
m = 5
        t                V              Vc
   (micro-sec)         (volt)         (volt)
   -----------------------------------------
        0.0             0.7000         0.6791
        4.0             0.9000         0.9377
        8.0             0.9000         0.9102
       12.0             0.7000         0.6748
       16.0             0.3000         0.3211
       20.0             0.0000        -0.0588
       24.0            -0.3000        -0.3790
       28.0            -0.7000        -0.5691
       32.0            -0.7000        -0.5833
       36.0            -0.3000        -0.4088
       40.0             0.0000        -0.0754
       44.0             0.3000         0.3361
       48.0             0.7000         0.6845
       52.0             0.7000         0.7592
       56.0             0.3000         0.2717


m          mse
--------------------
2          0.10789
3          0.10583
4          0.00803
5          0.00446
>>
```

See Figure 8.2.
---

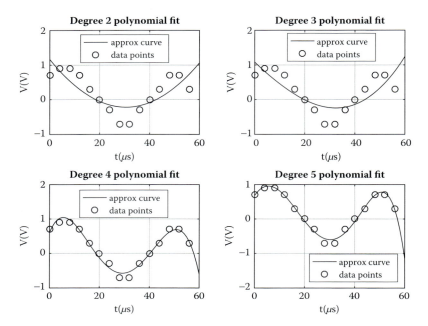

**FIGURE 8.2**
Approximating function and data points for audio signal versus time.

## 8.3 Curve Fitting with the Exponential Function

Many physical systems can be modeled as exponential functions. If your experimental data appears to fall into this category, it can be fitted with a function of the form

$$y_c = \beta_2 \, e^{-\beta_1 x} \tag{8.4}$$

where $\beta_1$ and $\beta_2$ are real constants.

Let us assume that a set of $n$ measured data points $(x_1, y_1), (x_2, y_2), \ldots, (x_n, y_n)$ exists. Then, let $z_i = \ln y_i$ and $z_c = \ln y_c = \ln \beta_2 - \beta_1 x$, and also let $a_1 = -\beta_1$ and $a_2 = \ln \beta_2$. Then taking the log of both sides of Equation 8.4 and making the above substitutions, we obtain the linear equation

$$z_c = a_1 x + a_2 \tag{8.5}$$

For the data points $(x_1, y_1), (x_2, y_2), \ldots, (x_n, y_n)$, the new set of data points becomes $(x_1, z_1), (x_2, z_2), \ldots, (x_n, z_n)$.

# Curve Fitting

We can then use MATLAB's polyfit function to determine $a_1$ and $a_2$. Then, $\beta_2 = e^{a_2}$ and $\beta_1 = -a_1$.

## Example 8.3

Suppose we took an oscilloscope picture of the position, $y$, of the mass in a mass-spring-dashpot system as shown in Figure 8.3 and measured the $(t, y_e)$ positions of the envelope. Table 8.1 gives the measured position, $y_e$, as a function of time, $t$. The governing equation of the envelope is

$$y_e = y_0\, e^{-\frac{c}{2m}t} \tag{8.6}$$

where:
  $c$ is the damping constant
  $m$ is the mass
  $y_e$ is the ordinate position of the envelope of the plot of the mass displacement from the equilibrium position

Comparing Equation 8.6 with Equation 8.4 we see that

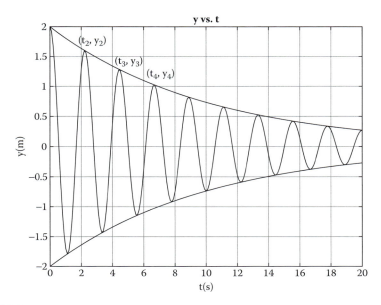

**FIGURE 8.3**
Position of a mass in a mass-spring-dashpot system with its envelope.

**TABLE 8.1**
$y_e$ versus $t$

| $t$ (s) | $y_e$ (cm) |
|---|---|
| 0.00 | 2.00 |
| 2.17 | 1.58 |
| 4.31 | 1.32 |
| 6.72 | 1.04 |
| 8.96 | 0.89 |
| 11.21 | 0.68 |
| 13.28 | 0.55 |
| 15.52 | 0.41 |
| 17.93 | 0.36 |

$$\beta_2 = y_0 \text{ and } \beta_1 = \frac{c}{2m} \text{ with } t \text{ replacing } x$$

Thus, the damping factor, $c$, for the system is given by

$$c = 2m\beta_1$$

Table 8.1 gives the measured values $y_e$ versus $t$.

As a check that the equation of the envelope is truly an exponential we can plot the data on semilog paper. It should plot as a straight line on semilog paper. This is done in the following program.

The mass, $m$, in the system is 25 kg.

**The Program Follows:**

```
% Example_8_3.m
% This program determines the best exponential fit for the envelope
% of the motion of a mass-spring-dashpot system.
clear; clc;
t=[0.00 2.17 4.31 6.72 8.96 11.21 13.28 15.52 17.93];
ye=[2.00 1.58 1.32 1.04 0.89 0.68 0.55 0.41 0.36];
z=[log(2.0) log(1.58) log(1.32) log(1.04) log(0.89) log(0.68) ...
    log(0.55) log(0.41) log(0.36)];
% If the relationship is exponential, ye should plot as a
% straight line on semi-log paper.
semilogy(t,ye,'x'), xlabel('t(s)'), ylabel('log(ye)'), grid,
title('log(ye) vs. t');
figure;
a=polyfit(t,z,1);
zc=a(1)*t+a(2);
```

# Curve Fitting

```
fprintf('a(1)=%7.3f a(2)=%6.3f \n',a(1),a(2));
beta(1)=-a(1);
beta(2)=exp(a(2));
fprintf('beta(1)=%7.3f, beta(2)=%6.3f \n',beta(1),beta(2));
for i=1:9
    yc(i)=beta(2)*exp(-beta(1)*t(i));
end
plot(t,yc,t,ye,'o'), xlabel('t(s)'), ylabel('yc(cm)'),
title('yc vs. t'), grid, legend('yc','ye');
m=25.0;
c=2*m*beta(1);
fprintf('m=%5.1f(kg), The damping constant=%7.4f(N-s/cm) \n',m,c);
```

**Program Results:**

```
a(1)= -0.097, a(2)= 0.695
beta(1)= 0.097, beta(2)= 2.005
m= 25.0(kg), The damping constant= 4.8718(N-s/cm)
>>
```

See Figures 8.4 and 8.5.

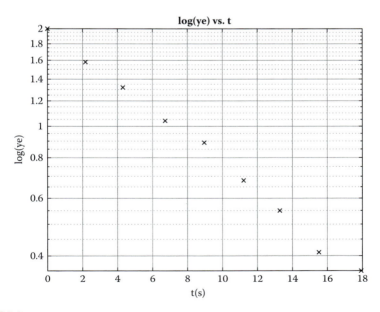

**FIGURE 8.4**
Plot of log($y_e$) versus t.

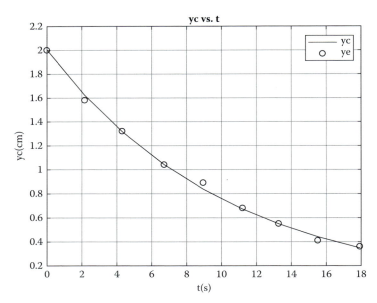

**FIGURE 8.5**
Plot of $y_c$ and $y_e$ versus t.

### REVIEW 8.1

1. Suppose an experiment produced a set of data and we wished to create an approximating curve, $y_c$, that is a polynomial expression that best fits the data. What is the name of the MATLAB function that will do this?
2. After executing MATLAB's `polyfit` function you may wish to obtain values on the approximating curve, $y_c$ at positions $(x_1, x_2, x_3, \ldots, x_n)$. If so, what MATLAB function would you use?

## 8.4 Cubic Splines

Suppose that we are given a set of $n$ data points and that we select an $m$th degree polynomial-approximating curve that produces curve values that are not allowed. For example, suppose it is known that a particular property represented by the approximating curve (such as absolute pressure or absolute temperature) must be positive and the approximating function produces values that are negative. In this case the approximating curve would not be satisfactory. The method of cubic splines eliminates this problem.

# Curve Fitting

Given a set of $(n + 1)$ data points $(x_i, y_i)$, $i = 1, 2, \ldots, (n + 1)$, the method of cubic splines develops a set of $n$ cubic functions, such that $y(x)$ is represented by a different cubic in each of the $n$ intervals and the set of cubics passes through all $(n + 1)$ data points.

## 8.4.1 MATLAB's Cubic Spline Curve-Fitting Function

The syntax for MATLAB's cubic `spline` function is

$$yy = \text{spline}(x_i, y_i, xx)$$

where $(x_i, y_i)$ is a given set of data points and $yy$ is the value of $y$ at $xx$. The `spline` function determines the four cubic coefficients for each section in the given data and will evaluate $yy$ by the cubic-spline method. The same result can be obtained by using MATLAB's `interp1` function and specifying the use of the spline method of interpolation. The syntax for interpolating by the `spline` method is

$$y_i = \text{interp1}(x, y, x_i, \text{'spline'})$$

### Example 8.4

The following example involves a measured increase in air pressure at distances from a blast. The data specifies the pressure above normal atmospheric pressure and is designated as overpressure. The program demonstrates the use of the MATLAB's spline function as well as MATLAB's interp1 function with the spline option to determine the pressure at distances not in the data. We see that the two methods produce the same results. The program follows:

```
% Example_8_4.m
% This program uses both MATLAB's spline function and MATLAB's
% interp1 function with the cubic spline option to determine the
% over-pressure resulting from a blast. The program calculates the
% over-pressure at locations between data points. The over-pressure
% is in kPa and the distance from the blast in km.
clear; clc;
dist=0.52:0.3:4.12;
press=[165.5 96.5 69.0 52.4 37.2 27.6 21.4 17.2 13.8 11.7 ...
       10.3 9.0 7.2];
d=0.52:0.1:4.12;
p1=spline(dist,press,d);
p2=interp1(dist,press,d,'spline');
fo=fopen('output.txt','w');
fprintf(fo,'Peak over-pressure vs. distance from the blast, \n');
fprintf(fo,'Cubic spline fit \n');
fprintf(fo,'dist(km)      over-press(kPa)      over-press(kPa)   \n');
fprintf(fo,'              by spline function   by interp1    \n');
fprintf(fo,'-----------------------------------------------------\n');
```

```
for n=1:length(d)
    fprintf(fo,'%5.2f  %10.2f   %10.2f  \n',d(n),p1(n),p2(n));
end
plot(d,p1,d,p2,'o'), xlabel('km from ground zero'),
ylabel('overpressure(kPa)'), grid,
title('Peak over-pressure vs. distance from blast')
fclose(fo);
```

**Program Results:**

```
Peak over-pressure vs. distance from blast, cubic spline fit
dist(km)    over-press(kPa)      over-press(kPa)
            by spline function    by interp1
-------------------------------------------------
0.52             165.50               165.50
0.62             135.72               135.72
0.72             113.15               113.15
0.82              96.50                96.50
0.92              84.46                84.46
1.02              75.72                75.72
1.12              69.00                69.00
1.22              63.15                63.15
1.32              57.71                57.71
1.42              52.40                52.40
1.52              47.02                47.02
  .                 .                    .
  .                 .                    .
  .                 .                    .
3.12              12.28                12.28
3.22              11.70                11.70
3.32              11.19                11.19
3.42              10.73                10.73
3.52              10.30                10.30
3.62               9.88                 9.88
3.72               9.46                 9.46
3.82               9.00                 9.00
3.92               8.49                 8.49
4.02               7.89                 7.89
4.12               7.20                 7.20
```

See Figure 8.6.

### Example 8.5

Example 8.2 can also be used as an example for the use of MATLAB's interp1 function with the spline option. That example involved the sampling with respect to time of an audio signal that is converted to a voltage by a microphone and an amplifier. The data is in volts (V) versus time in microseconds (μs). Variable names in this program differ from those in Example 8.2. The program follows:

# Curve Fitting

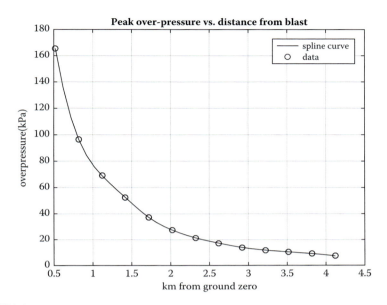

**FIGURE 8.6**
Blast overpressure versus distance from blast.

```
% Example_8_5.m
% This program uses interpolation by cubic splines to upsample
% an audio signal V(t) vs. time in microsec.
clear; clc;
% Define original data points for V(t) (time is in microsec)
orig_t = [ 0 4 8 12 16 20 24 28 32 36 40 44 48 52 56 ];
orig_V = [.7 .9 .9 .7 .3 0 -.3 -.7 -.7 -.3 0 .3 .7 .7 .3 ];
% Define upsampled time points
upsample_t = 0:60;
% Calculate interpolated data points using cubic splines
upsample_V = interp1(orig_t,orig_V,upsample_t,'spline');
% Print output to screen
fprintf('Upsampling via cubic spline fit \n');
fprintf('time (microsec) upsample_V \n');
for i=1:length(upsample_t)
    fprintf('%8.2f    %10.3f \n',upsample_t(i),upsample_V(i));
end
plot(orig_t,orig_V,'o',upsample_t,upsample_V);
xlabel('t(microsec)'); ylabel('V(volt)'); grid;
title('Upsampling with Cubic Spline Interpolation');
legend('original','upsampled');
```
--------------------------------------------------------------

**Program Results:**
Only the plot (Figure 8.7) is shown here.

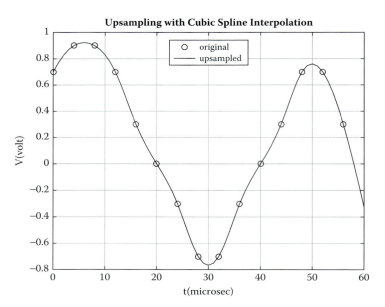

**FIGURE 8.7**
Upsampling of audio signal.

## Projects

**P8.1.** A formula describing the fluid level, $h$, in a tank as a function of time as the fluid discharges through a small circular orifice (see Figure P8.1) is

$$h = h_o - \sqrt{2 g h_o} \times \frac{C_d A_o}{A_T} t + \left(\frac{C_d A_o}{2 A_T}\right)^2 \times 2 g t^2 \qquad (P8.1)$$

where:
  $C_d$ is the discharge coefficient
  $h_o$ is the fluid level in the tank at time, $t = 0$
  $A_o$ is the area of the orifice
  $A_T$ is the cross-sectional area of the tank

An experiment consisting of a cylindrical tank with a small circular orifice was used to determine $C_d$ for that particular orifice and cylinder. The tank walls were transparent and a ruler was pasted to the wall allowing for the determination of the fluid level in the tank. The procedure was to fill the tank with water while the orifice was plugged. The plug was then removed and the water was allowed to flow through the orifice. The water level in the tank, $h_{exp}$ in m, was recorded as a function of time, $t$, in s. The experimental data is shown below:

# Curve Fitting

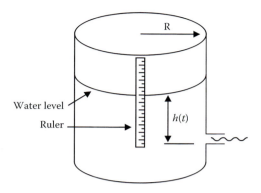

**FIGURE P8.1**
Water in a tank discharging through an orifice.

$$h_{exp} = [0.288\ 0.258\ 0.234\ 0.215\ 0.196\ 0.178\ 0.160\ 0.142\ 0.125\ 0.110\ \ldots$$
$$0.095\ 0.080\ 0.065\ 0.053\ 0.041\ 0.031\ 0.022\ 0.013\ 0.006\ 0.002\ \ldots$$
$$0.000]$$

$$t = [0\ 10\ 20\ 30\ 40\ 50\ 60\ 70\ 80\ 90\ 100\ 110\ 120\ 130\ 140\ 150\ 160\ 170\ \ldots$$
$$180\ 190\ 200]$$

Diameter of the orifice, $d_o = 0.0055$ m and the diameter of the tank, $D_T = 0.146$ m. The free surface elevation, $h_o$, at $t = 0$ is 0.288 m. The gravitational constant, $g = 9.81$ m/s².

Use the mse as defined by Equation P8.2 to determine the value for $C_d$ that best fits the data. Vary $C_d$ from 0.3 to 0.9 in steps of 0.01 and evaluate the mse for each $C_d$ selected, where

$$\text{mse} = \frac{1}{N}\sum_{i=1}^{N}[h(t_i) - h_{exp}(t_i)]^2 \qquad (P8.2)$$

where:
  $N$ is the number of data points
  $h(t_i)$ is the water level in the tank at $t_i$ as determined by Equation P8.1
  $h_{exp}(t_i)$ is the water level in the tank at $t_i$ as determined by experiment

For the $C_d$ with the lowest mse, create a plot of $h$ versus $t$ (solid line) and superimpose $h_{exp}$ versus $t$ as little x's onto the plot of $h$ versus $t$. Also print out the value of $C_d$ that gives the lowest mse.

**P8.2.** This project involves determining the best-fit polynomial approximating curve to the ($H$ vs. $Q$) data obtained experimentally. The experimental ($H$ vs. $Q$) data are shown in Table P8.1.

### TABLE P8.1
Experimental $H$ versus $Q$ Data

| $Q$ (m³/h) | $H$ (m) | $Q$ (m³/h) | $H$ (m) |
|---|---|---|---|
| 3.3 | 43.3 | 61.6 | 40.8 |
| 6.9 | 43.4 | 68.5 | 39.6 |
| 13.7 | 43.6 | 75.3 | 38.7 |
| 20.5 | 43.6 | 82.2 | 37.2 |
| 27.4 | 43.3 | 89 | 36.3 |
| 34.2 | 43.0 | 95.8 | 34.4 |
| 41.1 | 42.7 | 102.7 | 32.6 |

Try degree polynomials of two through four to determine which degree polynomial will give the smallest mse. Use MATLAB's function `polyfit` that returns the coefficients for each of the three polynomials. Then use the following MATLAB's function `polyval` to create for each polynomial:

1. A table containing $Q$, $H$, and $H_c$, where $H_c$ are values from the approximating curve for $H$ versus $Q$.
2. A plot containing both $H_c$ versus $Q$ (solid line) and $H$ versus $Q$ (small circles).

**P8.3.** Figure P8.2 shows a resistor-diode circuit using a type 1N914 silicon diode (D1) and a 10 kΩ resistor (R1). Table P8.2 shows a list of laboratory measurements of $v_2$ for various applied voltage levels of $v_1$ at room temperature (300 K). An equation that describes the behavior of the diode is given in Equation P8.3.

1. Use the technique described in Section 8.3 to find the best-fit values for $I_S$ and $v_T$.

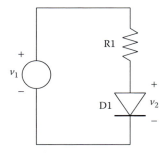

**FIGURE P8.2**
Diode-resistor circuit for laboratory measurement of diode I-$v$ curve.

# Curve Fitting

**TABLE P8.2**

Laboratory Measurements of Resistor-Diode Circuit

| $v_1$ (volts) | $v_2$ (volts) |
|---|---|
| 0.333 | 0.317 |
| 0.393 | 0.356 |
| 0.819 | 0.464 |
| 1.067 | 0.487 |
| 1.289 | 0.501 |
| 1.656 | 0.518 |
| 1.808 | 0.522 |
| 2.442 | 0.541 |
| 3.949 | 0.566 |
| 4.971 | 0.579 |
| 6.005 | 0.588 |
| 6.933 | 0.595 |
| 7.934 | 0.602 |
| 9.014 | 0.607 |
| 10.040 | 0.613 |
| 11.009 | 0.619 |
| 15.045 | 0.634 |
| 19.865 | 0.647 |
| 24.64 | 0.657 |
| 29.79 | 0.666 |

$$\frac{v_2 - v_1}{R_1} = I_S \exp\left(\frac{v_2}{v_T}\right) \qquad \text{(P8.3)}$$

Plot both the lab data and your fitted curve on the same axes.

**NOTE:** The diode current, $i_D = \dfrac{v_1 - v_2}{R_1}$.

2. $v_T = kT/q$ is known as the *thermal voltage* (where $k$ is the Boltzmann constant and $q$ is the unit electric charge). For your best-fit value for $v_T$, what is the corresponding temperature value $T$ (in Kelvin)? Take $k = 1.38\text{e}{-}23$, $q = 1.6\text{e}{-}19$.

# 9

# Numerical Integration

## 9.1 Introduction

In this chapter, we cover Simpson's rule for approximating the value of definite integrals as well as MATLAB®'s `integral` function. Understanding the concept in Simpson's rule will help you implement MATLAB's `integral` function for evaluating definite integrals. A discussion of MATLAB's `integral2` function is also included. Finally, examples demonstrating the usage of these three methods are given.

## 9.2 Numerical Integration and Simpson's Rule

We can evaluate a definite integral of a single variable using Simpson's rule. In applying Simpson's rule for evaluating $\int_A^B f(x)dx$, the first thing one does is subdividing the $x$ domain into $N$ intervals, where $N$ is an even number, giving $x_1, x_2, ..., x_N, x_{N+1}$. We then determine the functional values at the $x_n$ positions giving $f_1, f_2, ....... f_N, f_{N+1}$. We then connect three points on the curve $f(x)$ with second-degree polynomials (parabolas) and sum the areas under the parabolas to obtain the approximate area under the curve (see Figures 9.1 and 9.2). For a complete derivation see Article 6.3 in Reference 1. The final formula for the integral by Simpson's rule is

$$I = \int_{x_1}^{x_{N+1}} f(x)dx = \frac{\Delta x}{3}\left[f_1 + 4f_2 + 2f_3 + 4f_4 + 2f_5 + \cdots + 4f_N + f_{N+1}\right] \quad (9.1)$$

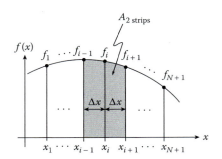

**FIGURE 9.1**
Area under two adjacent strips.

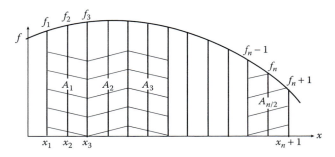

**FIGURE 9.2**
Summing all two-strip areas in Simpson's rule.

**Example 9.1**

Solve by Simpson's rule:

$$I = \int_0^{10} (x^3 + 3.2x^2 - 3.4x + 20.2)dx \tag{9.2}$$

```
% Example_9_1.m
% This program calculates an integral given in Equation 9.2 by
% Simpson's Rule.
% The integrand is: x^3+3.2*x^2-3.4*x+20.2
% The limits of integration are from 0-10.
clear; clc;
A=0; B=10;
N=100; dx=(B-A)/N;
% Compute values of x and f at each point:
% An arithmetic expression involving vector x produces a vector f.
% Need to use element by element multiplication.
x = A:dx:B;
```

```
f = x.^3+3.2*x.^2-3.4*x+20.2;
% Use two separate loops to sum up the even and odd terms
% of Simpson's Rule. Also, exclude endpoints in the loop, that is
% f(1) and f(N+1).
sum_even=0.0;
for i=2:2:N
    sum_even=sum_even+f(i);
end
sum_odd=0.0;
for i=3:2:N-1
    sum_odd=sum_odd+f(i);
end
% Calculate integral as per Equation 9.2.
I = dx/3 * (f(1) + 4*sum_even + 2*sum_odd + f(N+1));
% Display results
fprintf('Integrand: x^3+3.2*x^2-3.4*x+20.2 \n');
fprintf('Integration limits: %.1f to %.1f \n',A,B);
fprintf('Simpson rule solution, I = %9.4f \n',I);
% Compare with analytic solution.
% Analytic solution:
I2 = 0.25*10^4+3.2/3*10^3-3.4/2*10^2+20.2*10;
fprintf('Analytic solution, I2 = %9.4f \n',I2);
```
---

**Program Results:**
```
Integrand: x^3+3.2*x^2-3.4*x+20.2
Integration limits: 0.0 to 10.0
Simpson rule solution, I = 3598.6667
Analytical solution, I2 = 3598.6667
>>
```
---

We see that solving the integral of Example 6.1 by Simpson's rule gives the same answer as the analytical method up to four decimal places.

**Exercises**

**E9.1.** Evaluate the following definite integrals by Simpson's rule:

1. $I = \int_0^3 \dfrac{dx}{5e^{3x} + 2e^{-3x}}$

2. $I = \int_{-\pi/2}^{\pi/2} \dfrac{\sin x \, dx}{\sqrt{1 - 4\sin^2 x}}$

3. $I = \int_0^{\pi} (\sinh x - \cos x) \, dx$

## 9.3 Improper Integrals

An integral is improper if the integrand approaches infinity at some point within the limits of integration, including the end points. In many cases, the integration will still result in a finite solution. An example of an improper integral follows:

$$I = \int_0^1 \frac{\log(1+x)}{x} dx \tag{9.3}$$

The above integral is improper since both the numerator and denominator are zero at the lower limit ($x = 0$). The exact value of $I$ can be obtained by residue theory in complex variables and in this case the integral, $I$, evaluates to ($\pi^2/12$) = 0.822467. MATLAB's `integral` function, which is discussed next, is able to evaluate some improper integrals, but may give you a warning that the answer may be inaccurate.

## 9.4 MATLAB's `integral` Function

The MATLAB function for evaluating integrals is the function `integral`. It is a replacement for MALAB's `quad` function. A description of the function can be obtained by typing `help integral` in the Command Window. The syntax for MATLAB's `integral` function is

`Q = integral(FUN,A,B)`

where `FUN` is a function handle for the self-written function that describes the integrand. `A` and `B` are the limits of integration and `Q` is the integral result. The `integral` function approximates the integral using global adaptive quadrature and default error tolerances. Although the `integral` function can treat integrand variables that are complex, we are only interested for cases where the integrand involves only scalar value functions. The function will accept limits of integration `A` or `B` as inf or –inf. The function `Y=FUN(X)` should accept a vector argument `X` and return a vector result `Y`. The integrand is evaluated at each element of `X`. The function `FUN` can be either as a separate .m file or as an anonymous function. You may use the latter method if the integrand can be expressed in a single line.

The `integral` function is also able to evaluate certain improper integrals. It does this by selecting limits of integration that are very close to the singular points, but not on them, thus, removing the singularity.

## Example 9.2

We will now repeat Example 9.1, but this time we will use MATLAB's `integral` function to do the integration. The integral *I* in Example 9.1 is

$$I = \int_0^{10} (x^3 + 3.2x^2 - 3.4x + 20.2)dx$$

The program follows:

```
% Example_9_2.m
% This program evaluates the integral of the function 'f1'
% between A and B by MATLAB's integral function. Since the function 'f1'
% is just a single line, we can use the anonymous form of the function.
clear; clc;
f1=@(x) (x.^3+3.2*x.^2-3.4*x+20.2);
A=0.0; B=10.0;
I = integral(f1,A,B);
% Note f1 is not enclosed by single quotation marks.
fprintf('Integration of f1 over [%.0f,%.0f] ',A,B);
fprintf('by MATLAB''s integral function:\n');
fprintf('f1 = x^3+3.2*x^2-3.4*x+20.2 \n');
fprintf('integral = %10.4f \n',I);
```

**Program Results:**

```
Integration of f1 over [0,10] by MATLAB's integral function:
f1 = x^3+3.2*x^2-3.4*x+20.2
integral = 3598.6667
>>
```

We see that the results are the same as those obtained in Example 9.1.

## Example 9.3

$$\text{EVALUATE:} \quad I_2 = \int_0^1 \frac{t}{t^3+t+1}dt$$

```
% Example_9_3.m
% This program evaluates the integral of Example 9.3 by MATLAB's
% integral function. A separate .m file describes the integrand
% to be integrated.
% The integrand is t/(t^3+t+1.0)
clear; clc;
A=0.0; B=1.0;
I2 = integral(@f2_func,A,B);
fprintf('Integration of integrand over [%.0f,%.0f] ',A,B);
fprintf('by MATLAB''s integral function:\n');
fprintf('Integrand = t/(t^3+t+1) \n');
fprintf('integral=%f \n',I2);
```

```
% This function works with Example_9_3.m
function f=f2_func(t)
f = t./(t.^3+t+1.0);
```

**Program Results:**
```
Integration of integrand over [0,1] by MATLAB's integral function:
Integrand = t/(t^3+t+1)
integral=0.260069
>>
```

Let us evaluate the improper integral described by Equation 9.3 by MATLAB's `integral` function with limits from 0 to 1. Recall that the function $\log(1+x)/x$ is undefined at $x = 0$.

Again, we will use an anonymous function to describe the integrand.

**Example 9.4**

```
% Example_9_4.m
% This program evaluates the improper integral log(1+x)/x with
% limits from 0 to 1 using MATLAB's integral function.
clear; clc;
I3=@(x) log(1+x)./x;
A=0; B=1;
fprintf('This program uses MATLAB''s integral function to \n');
fprintf('evaluate the improper integral of log(1+x)/x \n');
fprintf('from %2.0f to %2.0f. \n',A,B);
I = integral(I3,A,B);
fprintf('I = %10.6f \n',I);
```

**Program Results:**
```
This program uses MATLAB's integral function to evaluate
the improper integral of log(1+x)/x from 0 to 1.
I =   0.822467
>>
```

We see that the answer by MATLAB's `integral` function is the same as shown in Section 9.3.

# Exercises

**E9.2.** Use MATLAB's `integral` function to evaluate the following integrals. Note that integral exercises 4, 5, and 6 are improper integrals.

1. $I = \displaystyle\int_0^3 \dfrac{dx}{5e^{3x} + 2e^{-3x}}$

2. $I = \displaystyle\int_{-\pi/2}^{\pi/2} \dfrac{\sin x\, dx}{\sqrt{1 - 4\sin^2 x}}$

3. $I = \displaystyle\int_0^{\pi} (\sinh x - \cos x)\, dx$

4. $I = \displaystyle\int_0^1 \dfrac{3e^x\, dx}{\sqrt{1 - x^2}}$

5. $I = \displaystyle\int_0^1 \dfrac{\log(1+x)\, dx}{(1-x)}$

6. $I = \displaystyle\int_0^1 \dfrac{\log(1+x)\, dx}{(1-x^2)}$

## REVIEW 9.1

1. What is the formula for evaluating the integral, $I = \int_A^B f(x)\, dx$ by the Simpson's rule?
2. What is the name of MATLAB's function for integrating a single variable function?
3. In MATLAB's function for integrating a single variable function how does one define the function to be integrated?
4. If the integrand contains nonlinear terms, how must they be treated?
5. Will MATLAB's `integral` function treat improper integrals?

## 9.5 MATLAB's `integral2` Function

The MATLAB's function for numerically evaluating a double integral is `integral2`. This function replaces MATLAB's `dblquad` function. A description of the function follows:

Q = integral2(FUN, XMIN, XMAX, YMIN, YMAX)

where Q is the result of the double integration, FUN(X,Y) is a function handle for the two-dimensional *self-written* integrand function. The limits of integration are XMIN, XMAX, YMIN(X), YMAX(X), where XMIN <= X <= XMAX and YMIN(X) <= Y <= YMAX(X). YMIN and YMAX may be either a scalar value or a function handle.

**The self-written function FUN(X,Y) should accept vectors X and Y and return a vector Z of values of the integrand. The X and Y input variables to function FUN comes from MATLAB's `integral2` function and the output vector Z from function FUN goes to MATLAB's `integral2` function. The output, Q, from the `integral2` function goes to the program that calls the `integral2` function.**

**Example 9.5**

Calculate the volume of a hemisphere of radius, $R$, by MATLAB's `integral2` function.

To find the volume, we define a differential volume element, $dV$, as follows:

$dV = \sqrt{R^2 - x^2 - y^2}\,dxdy$ (as shown in Figure 9.3) and double-integrate over the intervals $x = [-R, R]$ and $y = [-\text{sqrt}(R^2 - x^2), \text{sqrt}(R^2 - x^2)]$.

```
% Example_9_5.m
% This program calculates the volume of a hemisphere (with R=1)
% using MATLAB's integral2 function. The solution is compared with the
% known exact solution for the volume of a hemisphere.
clear; clc;
R = 1;
ymin=@(x) -sqrt(R^2-x.^2);
ymax=@(x) sqrt(R^2-x.^2);
funz=@(x,y) sqrt(R^2-x.^2-y.^2);
V = integral2(funz,-R,R,ymin,ymax);
V_exact = 2/3*pi*R^3;
% print results
fprintf('Volume V of a hemisphere of radius %.1f m \n',R);
fprintf('V by intgral2 = %.4f m^3\n',V);
fprintf('V exact = %.4f m^3\n',V_exact);
```

# Numerical Integration

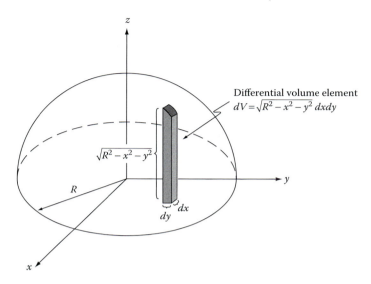

**FIGURE 9.3**
Infinitesimal volume inside a hemisphere.

## Program Results

```
Volume V of a hemisphere of radius 1.0 m
V by intgral2 = 2.0944 m^3
V exact = 2.0944 m^3
>>
```

---

### Example 9.6

The object shown in Figure 9.4 is enclosed by two curves, one of which is a straight line and the other is a parabola. The object thickness, $\Delta z$, is 5 cm. Take the object material to be steel with a mass density, $\rho = 8000$ kg/m³. The dimensions in the figure are also in cm.

1. Use MATLAB's `integral2` function to estimate the mass of the object.

   **NOTE:** $m = \rho \Delta z \iint_A dx\,dy$

   Print out $m$, include units.
2. Using 60 subdivisions on the $x$ domain, determine $y_{min}$ and $y_{max}$ for the region, where $y_{min}$ and $y_{max}$ are the minimum and maximum $y$ positions respectively in the region of interest. For every other $x$ position, print out a table of $x$, $y_{min}$, and $y_{max}$, include table headings and units.

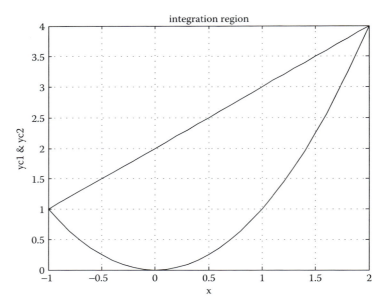

**FIGURE 9.4**
Object enclosed by two curves.

```
% Example_9_6.m
% This example determines the mass of an object that is enclosed
% by 2 curves. The lower curve is a parabola, y=x^2, and the upper
% curve is a straight line,y=x+2
% x range is from -1 to 2.
% y range is from 0 to 4.
% m=dz*Integration of (rho dx dy) from ymin to ymax.
% The input variables to FUN of MATLAB's integral2 function is (x,y).
% Since the integrand is 1, we needed to express the integrand as x./x
clear; clc;
rho=8e-3;
fun_9_6= @(x,y) x./x;
ymin=@(x) x.^2;
ymax=@(x) x+2;
dz=5.0;
m=dz*rho*integral2(fun_9_6,-1,2,ymin,ymax);
fprintf('m = %8.4f (kg) \n',m)
fprintf('\n');
x=-1:3/60:2;
ymin=x.^2;
ymax=x+2;
fprintf('  j      x(cm)      ymin(cm)      ymax(cm)    \n');
fprintf('-----------------------------------------\n');
for j=1:2:length(x)
    fprintf('%2i    %8.2f    %8.4f       %8.4f \n',...
            j,x(j),ymin(j),ymax(j));
end
```

```
plot(x,ymin,x,ymax), xlabel('x(cm)'),ylabel('ymin(cm) & ymax(cm)'),
grid, title('ymin & ymax vs. x');
```

**Program Results:**

m = 0.1800 (kg)

| j  | x(cm)  | ymin(cm) | ymax(cm) |
|----|--------|----------|----------|
| 1  | -1.00  | 1.0000   | 1.0000   |
| 3  | -0.90  | 0.8100   | 1.1000   |
| 5  | -0.80  | 0.6400   | 1.2000   |
| 7  | -0.70  | 0.4900   | 1.3000   |
| 9  | -0.60  | 0.3600   | 1.4000   |
| 11 | -0.50  | 0.2500   | 1.5000   |
| 13 | -0.40  | 0.1600   | 1.6000   |
| 15 | -0.30  | 0.0900   | 1.7000   |
| 17 | -0.20  | 0.0400   | 1.8000   |
| 19 | -0.10  | 0.0100   | 1.9000   |
| 21 | 0.00   | 0.0000   | 2.0000   |
| .  | .      | .        | .        |
| 51 | 1.50   | 2.2500   | 3.5000   |
| 53 | 1.60   | 2.5600   | 3.6000   |
| 55 | 1.70   | 2.8900   | 3.7000   |
| 57 | 1.80   | 3.2400   | 3.8000   |
| 59 | 1.90   | 3.6100   | 3.9000   |
| 61 | 2.00   | 4.0000   | 4.0000   |

\>\>

## REVIEW 9.2

1. What is the name of MATLAB's function for integrating a two-dimensional function?
2. List the arguments that go into MATLAB's function for integrating a two-dimensional function.

## Projects

**P9.1.** This exercise is from Thermodynamics. The entropy change of an ideal gas from state $(T_1, p_1)$ to state $(T_2, p_2)$ is given by

$$s(T_2, p_2) - s(T_1, p_1) = \int_{T_1}^{T_2} c_p(T) \frac{dT}{T} - R \ln \frac{p_2}{p_1} \qquad (P9.1)$$

where:
  $s$ is the entropy (kJ/kg-K)
  $c_p$ is the specific heat at constant pressure (kJ/kg-K)

$p$ is the pressure (kPa)
$T$ is the absolute temperatue (K)
$R$ is the gas constant (kJ/kg-K)

The specific heat, $c_p(T)$, can be approximated by a fourth-degree polynomial [2], that is,

$$c_p(T) = R(a_1 + a_2 T + a_3 T^2 + a_4 T^3 + a_5 T^4) \tag{P9.2}$$

$$R = \frac{\bar{R}}{M}$$

where:
$\bar{R}$ is the Universal gas constant = 8.314 (kJ/kmol-K)
$M$ is the Molal mass (kg/kmol)

For carbon dioxide [2],

$$a_1 = 2.401, \; a_2 = 8.735 \times 10^{-3}, \; a_3 = -6.607 \times 10^{-6}, \; a_4 = 2.002 \times 10^{-9}, \; a_5 = 0.0$$

$$M = 44.01 \text{ kg/kmol}$$

Use MATLAB's `integral` function to calculate the change in entropy, $s(T_2, p_2) - s(T_1, p_1)$ for $(T_1, p_1) = (400 \text{ K}, 1.0 \text{ atm})$, $(T_2, p_2) = (900 \text{ K}, 10.0 \text{ atm})$. Print the results to the screen.

**NOTE:** 1 atm = $1.0132 \times 10^5 \text{ N/m}^2$

**P9.2.** An ice slab, initially at temperature, $T_i = -20°C$ is suddenly subjected to a change in air temperature, $T_\infty = 10°C$. This results in a heat transfer, $q$, per unit surface area from the air to the ice slab. An approximate formula for $q$ (J/m²) follows:

$$q = h(T_\infty - T_i) \int_0^{t_f} \left( e^{\left(\frac{h^2 \alpha t}{k^2}\right)} \times \left[ 1 - \text{erf}\left(\frac{h\sqrt{\alpha t}}{k}\right) \right] \right) dt \tag{P9.3}$$

where:
$k$ is the thermal conductivity of the slab material
$h$ is the convective heat transfer coefficient
$\alpha$ is the thermal diffusivity of the slab material
$T_\infty$ is the air temperature
$T_i$ is the initial slab temperature

**NOTE 1:** The error function, erf(x) is written in MATLAB as `erf(x)`.

**NOTE 2:** If $y = \text{erf}(x)$ and $x$ is a vector, then $y$ will also be a vector.

Assume: $k = 2.2$ W/m-C, $\alpha = 12.6 \times 10^{-7}$ m²/s, $h = 100$ W/m²-C, and $t_f = 792$ s.

# Numerical Integration

Develop a MATLAB program using MATLAB's `integral` function to evaluate $q$. Use a separate .m file to describe the integrand. Print the constants, $k$, $h$, $\alpha$, $T_i$, $T_\infty$ and the result, $q$ to the screen. Use e format for $q$. Include dimensions.

**P9.3.** We wish to determine the *x*-component of the Electric field at position $(x_o, y_o, z_o)$ due to a line of point charges extending along the *z*-axis from $z = -0.01$ m to $z = +0.01$ m (see Figure P9.1). For the derivation of the governing equations for the electric field see Project P6.3 in [1]. Assume that $dQ = \lambda dz_p$ with $\lambda = 2 \times 10^{-9}$ C/m. Here $dQ$ is the strength of the point charge distribution. The *x*-component of the electric field at position $(x_o, y_o, z_o)$ is given by

$$E_x(x_o, y_o, z_o) = \int_{-0.01}^{0.01} \frac{\lambda dz_p}{4\pi\varepsilon_o} \frac{x_o - x_p}{\left((x_o - x_p)^2 + (y_o - y_p)^2 + (z_o - z_p)^2\right)^{3/2}} \quad (P9.4)$$

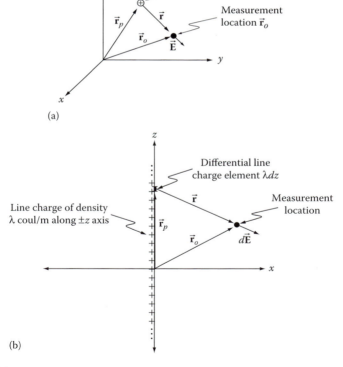

**FIGURE P9.1**
(a) Electric field at $(x_o, y_o, z_o)$ due to point charge at $(x_p, y_p, z_p)$. (b) Electric field at $(x_o, y_o, z_o)$ due to line charge element $\lambda dz$ located along the *z*-axis.

where $(x_p, y_p, z_p)$ is the coordinates of the point charge. Use MATLAB's integral function to determine $E_x(0.005, 0, 0)$. The units of $E_x$ is V/m. Take $\varepsilon_o = 8.85 \times 10^{-12}$ farad/m.

**P9.4.** The solution for the displacement, $Y(x,t)$, from the horizontal of a vibrating string (see Figure P9.2) is given by

$$Y(x,t) = \sum_{n=1}^{\infty} a_n \sin \frac{n\pi x}{L} \cos \frac{n\pi ct}{L} \tag{P9.5}$$

where

$$a_n = \frac{2}{L} \int_0^L f(x) \sin \frac{n\pi x}{L} dx \tag{P9.6}$$

and

$$f(x) = Y(x, 0) \tag{P9.7}$$

For a compete derivation of the governing equation see Section 13.2 in Reference 1.

Use MATLAB's integral function to determine $a_n$, for $n = 1, 2, \ldots, 10$. Create a table and a plot of $a_n$ versus $n$. Take $L = 1.0$ m and

$$f(x) = \begin{cases} 0.4x, & 0 \leq x \leq 0.75L \\ 1.2 - 1.2x, & 0.75L \leq x \leq L \end{cases} \tag{P9.8}$$

**P9.5.** The components of the electric field $E_x, E_y,$ and $E_z$ resulting from a line of point charges with a linear charge density, $\lambda$, with units Coulomb/meter, which is evenly distributed along the z-axis from $z = -0.01$ m to $z = +0.01$ m (see Figure P9.1) is given by

$$E_x(x_o, y_o, z_o) = \int_{-0.01}^{0.01} \frac{\lambda dz_p}{4\pi\varepsilon_o} \frac{x_o - x_p}{\left((x_o - x_p)^2 + (y_o - y_p)^2 + (z_o - z_p)^2\right)^{3/2}} \tag{P9.9}$$

$$E_y(x_o, y_o, z_o) = \int_{-0.01}^{0.01} \frac{\lambda dz_p}{4\pi\varepsilon_o} \frac{y_o - y_p}{\left((x_o - x_p)^2 + (y_o - y_p)^2 + (z_o - z_p)^2\right)^{3/2}} \tag{P9.10}$$

**FIGURE P9.2**
Vibrating string.

# Numerical Integration

**TABLE P9.1**

Table Format for Presenting $E_x$ Values

| x | $E_x$ Values y | | | | | |
|---|---|---|---|---|---|---|
| | −0.05 | −0.03 | −0.01 | 0.01 | 0.03 | 0.05 |
| −0.05 | — | — | — | — | — | — |
| −0.04 | — | — | — | — | — | — |
| −0.03 | — | — | — | — | — | — |
| . | | | | | | |
| . | | | | | | |
| 0.03 | — | — | — | — | — | — |
| 0.04 | — | — | — | — | — | — |
| 0.05 | — | — | — | — | — | — |

$$E_z(x_o, y_o, z_o) = \int_{-0.01}^{0.01} \frac{\lambda dz_p}{4\pi\varepsilon_o} \frac{z_o - z_p}{\left((x_o - x_p)^2 + (y_o - y_p)^2 + (z_o - z_p)^2\right)^{3/2}} \quad \text{(P9.11)}$$

where:
$(x_p, y_p, z_p)$ represents the position of the point charges
$(x_o, y_o, z_o)$ represents a point of interest in the vicinity of the point charges

The units of $E_x, E_y$, and $E_z$ are Newton/Coulomb (N/C) or Volt/m (V/m).
Take $\varepsilon_o = 8.85 \times 10^{-12}$ farad/m.

Create a MATLAB program that will determine the electric field component, $E_x$ and $E_y$ in the $(x, y)$ plane for the interval $-50 \leq x \leq 50$ mm and $-50 \leq y \leq 50$ mm with a step size of 10 mm. Omit the point $(x, y) = (0,0)$. Due to symmetry, assume that $E_z = 0$. Print $E_x$ and $E_y$ in separate tables using a table format as shown in Table P9.1. Print $E_x$ and $E_y$ to one decimal place.

**P9.6.** This project involves determining the surface area of a hemisphere of radius one meter.

A differential surface area on the hemisphere is $dA = (R\sin\phi d\theta) Rd\phi$ as shown in Figure P9.3. Create a MATLAB program using MATLAB's integral2 function to find the surface area. Take $\phi = [0, \pi/2]$ and $\theta = [0, 2\pi]$. Compare your answer with the known exact expression for the surface area of a hemisphere, which is $2\pi R^2$.

**P9.7.** An object is enclosed by two curves, one of which is a straight line and the other is a parabola. The $x$ range of the object is from −2 cm to +4 cm. The equation of the parabola is

$$y = 6 - 1.5x^2$$

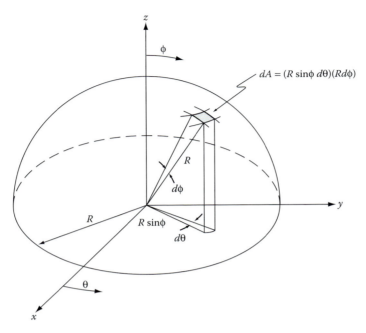

**FIGURE P9.3**
A differential surface area on the hemisphere.

The equation of the straight line is

$$y = -3x - 6$$

The object thickness, $\Delta z$, is 5 cm. Take the object material to be steel with a mass density, $\rho = 8000$ kg/m³.

a. Using 60 subdivisions on the $x$ domain, determine $y_{min}$ and $y_{max}$ for the region, where $y_{min}$ and $y_{max}$ are the minimum and maximum $y$ positions respectively in the region of interest. For every other $x$ position, print out a table of $x$, $y_{min}$ and $y_{max}$, include table headings and units.
b. Create a two dimensional plot of the object.
c. Use MATLAB's integral2 function to estimate the mass of the object.

Print out to the Command Window the mass, $m$, include units.

**NOTE:** $m = \Delta z \iint_A \rho \, dx dy.$

# Numerical Integration

**P9.8.** The $(x_c, y_c)$ position of the center of mass of the object described in Project P9.7 is given by

$$mx_c = \Delta z \rho \iint_A x\,dxdy \qquad my_c = \Delta z \rho \iint_A y\,dxdy \qquad \text{(P9.12)}$$

where:
$\rho$ is the mass density of the material
$m$ is the mass of the object
$\Delta z$ is the thickness of the object

Create a MATLAB program that will evaluate $(x_c, y_c)$ using MATLAB's `integral2` function. Print the results to the screen to four decimal places.

**P9.9.** Using the infinitesimal volume shown in Figure 9.3 and MATLAB's `integral2` function, determine the centroid position, $z_c$, of the hemisphere described in Example 9.5. By symmetry, we can assume that $x_c = 0$ and $y_c = 0$. Noting that $z_c$ for the infinitesimal volume is at the center position, that is,

$$z_c\,dV = \frac{1}{2}\sqrt{R^2 - x^2 - y^2} \times \sqrt{R^2 - x^2 - y^2}\,dxdy$$

$$z_c V = \iint_A \frac{1}{2}(R^2 - x^2 - y^2)\,dxdy \qquad \text{(P9.13)}$$

where:
$x = [-R, R]$ and $y = [-\text{sqrt}(R^2 - x^2), \text{sqrt}(R^2 - x^2)]$
$V$ is the volume of the hemisphere $= 2/3\pi R^3$

We can also determine the centroid position analytically by taking an infinitesimal volume shown in Figure P9.4, then

$$dV = \pi r^2 dz \qquad \text{(P9.14)}$$

where:

$$r = \sqrt{(R^2 - z^2)} \qquad \text{(P9.15)}$$

Thus,

$$z_c = \frac{\pi}{V}\int_0^R (R^2 z - z^3)\,dz = \frac{\pi}{\frac{2}{3}\pi R^3} \times \left[R^2\frac{z^2}{2} - \frac{z^4}{4}\right]_0^R = \frac{3}{8}R \qquad \text{(P9.16)}$$

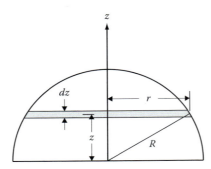

**FIGURE P9.4**
Infinitesimal region used to determine the centroid position $z_c$ analytically.

Compare $z_c$ obtained by the use of MATLAB's `integral2` function with the exact solution. Take $R = 1$.

## Reference

1. Bober, W., *Introduction to Numerical and Analytical Methods with MATLAB for Engineers and Scientists*, CRC Press, Boca Raton, FL, 2014.

# 10

# Numerical Integration of Ordinary Differential Equations

## 10.1 Introduction

Many ordinary differential equations (ODE) result from a particular physical law. The physical law is a mathematical model of some particular physical phenomenon. Many of the equations that have been used in this book are based on Newton's second law of motion. For example, the equations used to describe the motion of a free falling ball in a gravitational field (Example 2.7) or the motion of the mass in a mass-spring-dashpot system (Exercise E2.3-for a complete derivation of the governing equations, see Project P2.5 in Reference 1), or the velocity and position of the basketball (Exercise E2.4) are differential equations based on Newton's second law. The voltage in a parallel RLC circuit (Exercise E2.6) resulted from several electrical laws, including Kirchhoff's current law, which resulted in an ordinary differential equation whose solution is given in Equation 2.13 (for a complete derivation of the governing equations, see P2.7 in Reference 1). Ordinary differential equations can be broken up into two categories:

1. *Initial value problems* are those in which the initial conditions of the variables are known. All of the examples and exercises mentioned above fall into this category. Additional examples include launching a rocket with a known initial position and velocity or the value of a circuit node voltage (or its slope) at $t = 0$. In this chapter, we only cover the initial value problem along with MATLAB®'s built-in `ode45` function to solve these types of problems.

2. *Boundary value problems* in which we know variable conditions at specific coordinates in the problem geometry. For example, determining the temperature at various positions along a bar when the end temperatures at the bar ends are known. Other examples include determining the deflection of a beam along its length when the deflection at its ends is known, or determining the electric potential along the length of a conductor when the electric potential at both ends of a conductor are known. This topic is covered in Chapter 11.

## 10.2 Initial Value Problem and MATLAB's Ordinary Differential Equations Function

MATLAB has several built-in ODE functions that solve a system of first-order ordinary differential equations, including `ode23` and `ode45`. In this chapter, we will demonstrate MATLAB's `ode45` function, which is based on fourth- and fifth-order Runge–Kutta methods. A description of the `ode45` function follows (MATLAB's description of ode45 can be obtained by typing *help ode45* in the Command Window):

The ode45 function solves a system of first-order ordinary differential equations of the form $y'_n = f(t, y_1, y_2, \ldots, y_n)$ from time TO to TFINAL with initial conditions Y0. Here we have assumed that the independent variable is time, $t$, and the dependent variables are $y_1\ y_2\ \ldots y_n$.

The syntax for MATLAB's ode45 function is

[TOUT, YOUT] = ODE45(ODEFUN, TSPAN, Y0)

The ODEFUN argument is a function handle to the function describing the system of differential equations. TSPAN = [T0 Tfinal] is vector describing a time interval covered by the system of differential equations. Y0 is a vector describing the initial conditions. The function ODEFUN must take two input arguments: a scalar for the independent variable, $t$, and a vector for the dependent variables, $Y = [y_1\ y_2\ \ldots y_n]$. The output of function ODEFUN must be the system of differential equations as a column vector of the form

$$y'(1) = f_1(y(1), y(2), \ldots, y(n)),$$
$$y'(2) = f_2(y(1), y(2), \ldots y(n)),$$

.

.

$$y'(n) = f_2(y(1), y(2), \ldots y(n))$$

The time interval TSPAN is typically a two-element vector containing a start and end time; ode45 will automatically choose an appropriate time step (and might even vary the time step within the interval). ode45 will return two vectors: a list of time points TOUT and the solution YOUT at each time point. If you want to force ode45 to solve the system at specific time points, then you can explicitly specify the time points in TSPAN = [T0 T1.. TFINAL]. The output variable TOUT is a column vector equal in size to tspan and YOUT are column vectors of $y(1), y(2), \ldots y(n)$.

# Numerical Integration of Ordinary Differential Equations 207

**Example 10.1**

Let us consider the ball-bearing problem of Example E2.5. Applying Newton's second law to the ball bearing gives the following first-order differential equation:

$$\frac{W}{g}\frac{dV}{dt} = 6\pi R\mu(V_T - V) \qquad (10.1)$$

where:
    V is the ball-bearing velocity
    $V_T$ terminal velocity of the ball bearing $= (W-B)/6\pi R\mu$
    W is the weight of the ball bearing $= \rho_{steel}\, \upsilon g$
    B is the buoyancy acting on the ball bearing $= \rho_{fluid}\, \upsilon g$
    R is the radius of the ball bearing
    $\upsilon$ is the volume of the sphere $= (4/3)\pi R^3$
    $\rho$ is the mass density
    $\mu$ is the viscosity of the fluid
    g is the gravitational constant = 9.81 m/s²

To see the full derivation of Equation 10.1 see Exercise E2.5 in Reference 1.
    In the notation of ode45, $(dV/dt)$ = V'
    Take $\mu$ = 3.85 (N-s)/m², R = 0.01 m, $\rho_{steel}$ = 7910 kg/m³, $\rho_{oil}$ = 899 kg/m³.
    We will take V(0) = 0.

The program follows.

```
% Example_10_1.m
% This program determines the velocity of a ball bearing
% dropped in a vat of fluid. The ball bearing reaches a
% terminal velocity when the unbalanced force acting
% on the object is zero.
% The program compares the velocity determined by the
% by both an analytical solution and MATLAB's ode45 function.
clear; clc;
global R mu g VT W VT
R=0.01; rho_steel=7910; rho_fluid=899; mu=3.85; g=9.81;
vol=4/3*R^3;
W=rho_steel*g*vol;
B=rho_fluid*g*vol;
VT=(W-B)/(6*pi*R*mu);
Vo=0;
tspan=0:0.01:0.2;
[t,V]=ode45('dVdt',tspan,Vo);
% Closed form solution is V2
arg=6*pi*R*mu*g/W;
V2=VT*(1-exp(-arg*t));
fprintf('t(s)      V(m/s)      V2(m/s)   \n');
fprintf('---------------------------------\n');
for i=1:length(t)
    fprintf('%4.2f      %6.4f      %6.4f \n',t(i),V(i),V2(i));
end
```

```
fprintf('Terminal Velocity, VT= %6.4f(m/s) \n', VT);
plot(t,V,t,V2,'x'), xlabel('t(s)'), ylabel('V(m/s)'), grid,
title('V vs. t');
```
---
```
% dVdt.m
% This function works with Example_10_1.m
function Vprime=dVdt(t,V)
global R mu g VT W VT
Vprime= 6*pi*R*mu*g/W*(VT-V);
```
---

**Program Results:**

See Figure 10.1.

| t(s) | V(m/s) | V2(m/s) |
|------|--------|---------|
| 0.00 | 0.0000 | 0.0000 |
| 0.01 | 0.0629 | 0.0629 |
| 0.02 | 0.0945 | 0.0945 |
| 0.03 | 0.1103 | 0.1103 |
| 0.04 | 0.1183 | 0.1183 |
| 0.05 | 0.1223 | 0.1223 |
| 0.06 | 0.1243 | 0.1243 |
| 0.07 | 0.1253 | 0.1253 |
| 0.08 | 0.1259 | 0.1259 |
| 0.09 | 0.1261 | 0.1261 |
| 0.10 | 0.1262 | 0.1262 |
| .    | .      | .       |
| .    | .      | .       |

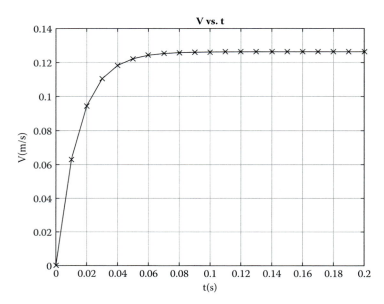

**FIGURE 10.1**
Velocity of ball bearing versus time.

```
0.15       0.1264          0.1264
0.16       0.1264          0.1264
0.17       0.1264          0.1264
0.18       0.1264          0.1264
0.19       0.1264          0.1264
0.20       0.1264          0.1264
Terminal Velocity, VT= 0.1264(m/s)
>>
```
---

**Example 10.2**

Whenever we deal with a second-order differential equation, we need to reduce the second-order differential equation to two first-order differential equations. Suppose we consider the mass motion in a mass-spring-dashpot system of Exercise E2.2. The governing differential equation is

$$y'' + \frac{c}{m}y' + \frac{k}{m}y = 0 \tag{10.2}$$

To see the full derivation of Equation 10.2 see Project P2.5 in Reference 1.
 To reduce Equation 10.2 to two first-order differential equations, let $y' = V$, then

$$V' = -\frac{c}{m}V - \frac{k}{m}y$$

$$y' = V \tag{10.3}$$

We will take $y(0) = 0.5$ and $y'(0) = 0$.
 We can now use MATLAB's ode45 to solve the system.
 The program follows:

```
% Example_10_2.m
% This program determines the position and velocity
% of a mass in a mass-spring-dashpot system using
% MATLAB's ode45 function.
% m=25 kg; c=5 N-s/m; k=100 N/m;
% Y(1)=y
% Y(2)=V
% Y(1)_prime=Y(2)
% Y(2)_prime= -c/m*Y(2)-k/m*Y(1)
clear; clc;
initial=[0.5 0.0];
tspan=0.0:0.1:10.0;
[t,Y]=ode45(@dYdt,tspan,initial);
y=Y(:,1);
V=Y(:,2);
fprintf(' t(s)       y(s)     V(m/s)   \n');
fprintf('-------------------------------------\n');
for i=1:2:101
    fprintf('%5.2f      %10.4f     %10.4f \n',t(i),y(i),V(i))
end
```

```
plot(t,y), xlabel('t(s)'), ylabel('y(m)'), grid,
title('y vs. t');
figure;
plot(t,V), xlabel('t(s)'), ylabel('V(m/s)'), grid,
title('V vs. t');
```
---
```
% dYdt.m
% This function works with Example_10_2.m
function Yprime=dYdt(t,Y)
m=25; c=5; k=100;
% Y(1)= y; Y(2)=V
Yprime=zeros(2,1);
Yprime(1)=Y(2);
Yprime(2)=-c/m*Y(2)-k/m*Y(1);
```
---

**Program Results:**

See Figure 10.2a and b.

| t(s) | y(m) | V(m/s) |
|---|---|---|
| 0.00 | 0.5000 | 0.0000 |
| 0.20 | 0.4611 | -0.3818 |
| 0.40 | 0.3523 | -0.6894 |
| 0.60 | 0.1932 | -0.8785 |
| 0.80 | 0.0105 | -0.9239 |
| 1.00 | -0.1668 | -0.8248 |
| 1.20 | -0.3111 | -0.6016 |
| 1.40 | -0.4017 | -0.2942 |
| 1.60 | -0.4265 | 0.0466 |
| 1.80 | -0.3845 | 0.3670 |
| 2.00 | -0.2844 | 0.6178 |
| . | . | . |
| 9.00 | 0.1231 | 0.3109 |
| 9.20 | 0.1729 | 0.1808 |
| 9.40 | 0.1939 | 0.0278 |
| 9.60 | 0.1841 | -0.1235 |
| 9.80 | 0.1462 | -0.2497 |
| 10.00 | 0.0871 | -0.3323 |

\>\>

---

# Numerical Integration of Ordinary Differential Equations

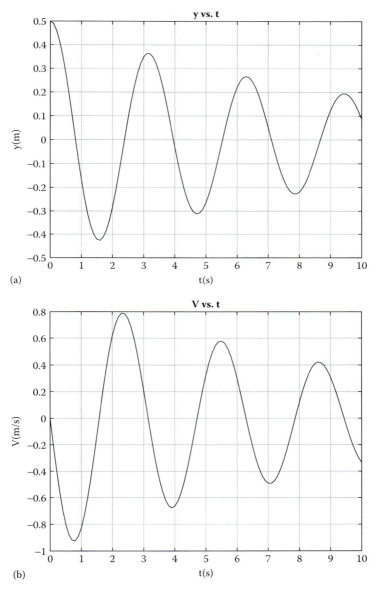

**FIGURE 10.2**
(a) Plot of mass displacement versus time and (b) plot of mass velocity versus time.

## Example 10.3

In this example, we consider the voltage in a parallel RLC circuit described in Project P2.10 (see Figure 2.22). For a complete derivation of the governing equation, see Project 2.7 in Reference 1. The governing differential equation for the circuit voltage is

$$\frac{d^2v}{dt^2} + \frac{1}{RC}\frac{dv}{dt} + \frac{1}{LC}v = 0 \tag{10.4}$$

To reduce Equation 10.4 to two first-order differential equations, let $v' = u$, then

$$u' = -\frac{1}{RC}u - \frac{1}{LC}v \tag{10.5}$$
$$v' = u$$

We will take $v(0) = 6$ and $\dfrac{dv}{dt}(0) = \dfrac{6}{RC}$

The program follows:

```
% Example_10_3.m
% This program determines the voltage in a parallel RLC circuit using
% MATLAB's ode45 function.
% R=100 ohm; L=1 mHc; C=1 microfarad;
% v(0)=6 volt; dvdt=v(0)/(R*C)
% Y(1)=v
% Y(2)=dv/dt=u
% Yprime(1)=Y(2)
% Yprime(2)= -1/(R*C)*Y(2)-1/(L*C)*Y(1)
clear; clc;
global R L C;
R=100; L=1e-3; C=1e-6;
initial=[6 6/(R*C)];
tspan=0:5e-6:5e-4;
[t,Y]=ode45(@dvoltdt,tspan,initial);
v=Y(:,1);
u=Y(:,2);
t2=t*1.0e+6;
fprintf('        t              v           dv/dt  \n');
fprintf('(micro-sec)         (volt)      (volt/sec)   \n');
fprintf('-------------------------------------------\n');
for i=1:length(t2)
    fprintf('    %5.0f           %10.2f      %10.0f \n',t2(i),v(i),u(i))
end
plot(t,v), xlabel('t(s)'), ylabel('v(volt)'), grid,
title('v vs. t');
figure;
plot(t,u), xlabel('t(s)'), ylabel('dv/dt(volt/s)'), grid,
title('dv/dt vs. t');
```
-----------------------------------------------------------------

# Numerical Integration of Ordinary Differential Equations

```
% This function works with Example_10_3.m
function Yprime=dvoltdt(t,Y)
global R L C;
% Y(1)=v
% Y(2)=dv/dt=u
Yprime=zeros(2,1);
Yprime(1)=Y(2);
Yprime(2)= -1/(R*C)*Y(2)-1/(L*C)*Y(1);
```
---

**Program Results:**

See Figure 10.3a and b.

```
       t              v            dv/dt
  (micro-sec)       (volt)       (volt/sec)
  ------------------------------------------
         0           6.00          60000
         5           6.22          27209
        10           6.27          -4644
        15           6.17         -34837
        20           5.93         -62711
        25           5.55         -87688
        30           5.06        -109310
        35           4.46        -127231
        40           3.79        -141188
        45           3.06        -151018
        50           2.29        -156678
         .            .              .
         .            .              .
       450           0.36         -20617
       455           0.25         -21094
       460           0.15         -21033
       465           0.04         -20461
       470          -0.06         -19418
       475          -0.15         -17955
       480          -0.24         -16127
       485          -0.31         -13996
       490          -0.38         -11629
       495          -0.43          -9094
       500          -0.47          -6462
>>
```
---

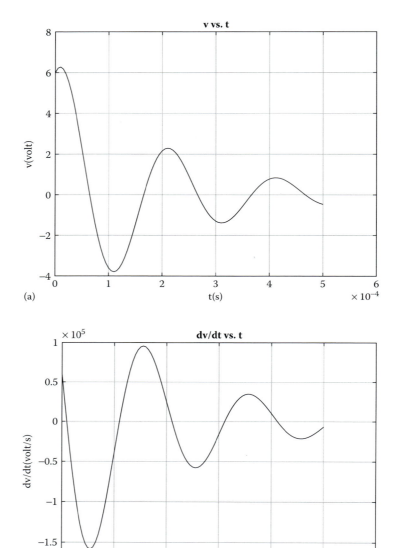

**FIGURE 10.3**
(a) Plot of voltage versus time. (b) Plot of dv/dt versus time.

# Numerical Integration of Ordinary Differential Equations 215

## Exercises

**E10.1.** The governing equation for the mass displacement in a mass-spring-dashpot system subjected to an oscillatory driving force is

$$y'' + \frac{c}{m}y' + \frac{k}{m}y = \frac{F_o}{m}\sin \omega t \tag{10.6}$$

Create a MATLAB program, using MATLAB's ode45 function to solve for $y$ and $y'$ as a function of time, $t$. The natural frequency of the system, $\omega_n = \sqrt{k/m}$. Let us consider two cases:

1. $\omega = 1.5\, \omega_n$
2. $\omega = \omega_n$

Take $m = 25$ kg, $c = 5$ N-s/m; $k = 100$ N/m; $F_o = 50$ N, $y(0) = 0.5$ m, $y'(0) = 0$.

Create a MATLAB program that uses MATLAB's ode45 function to solve for $y(t)$ for cases (1) and (2). Notice that when $\omega = 1.5\, \omega_n$ the amplitude of the oscillation grows with time and is much larger than the amplitude for the case when $\omega = \omega_n$. This is the effect of resonance.

**E10.2.** Solve the following system of three first-order differential equations using MATLAB's ode45 function:

$$y_1' = y_2 y_3 t$$

$$y_2' = -y_1 y_3$$

$$y_3' = -0.51 y_1 y_2$$

Initial conditions: $y_1(0) = 0$, $y_2(0) = 1.0$, and $y_3(0) = 1.0$.

**E10.3.** Solve the parallel RLC circuit of Example 10.3 for voltage, $v$, and the inductor current, $i_L$, by using MATLAB's ode45 function. The governing equations are

$$\frac{dv}{dt} = -\frac{1}{RC}v - \frac{1}{C}i_L$$

$$\frac{di_L}{dt} = \frac{1}{L}v$$

Assume $R = 50\,\Omega$, $L = 1\,\mu H$, $C = 10\,nF$, $v(0) = 3.3\,V$, $i_L(0) = 0\,A$.
Plot $v$ on $i_L$ on two separate graphs. Take $0 \le t \le 4$ μsec in steps of 0.01e-6 sec.

## Projects

**P10.1.** This project involves determining the temperature of a small solid aluminum sphere dropped into a fluid contained in a vertical circular cylinder of radius R. The sphere radius is $r$ and the fluid depth is $L$. Neglecting heat transfer to the container walls, the governing equations for this problem are

$$\left(mc\frac{dT}{dt}\right)_{al} = hA_s(T_f - T_{al}) \qquad \text{(P10.1a)}$$

$$\left(mc\frac{dT}{dt}\right)_f = hA_s(T_{al} - T_f) \qquad \text{(P10.1b)}$$

Use the following parameters for the problem:

$\rho_{al} = 2707 \text{ kg/m}^3$, $\rho_f = 880 \text{ kg/m}^3$, $c_{al} = 896 \text{ J/kg-°C}$, $c_f = 1880 \text{ J/kg-°C}$,

$T_{al}(0) = 80°C$, $T_f(0) = 20°C$, $r = 0.2$ m, $R = 0.3$ m, $L = 0.5$ m, $h = 890 \text{ W/m}^2\text{-°C}$

$0 \le t \le 300$ s in steps of 0.5 s.

Create a MATLAB program using MATLAB's `ode45` function to determine the temperatures of the aluminum sphere and the fluid. Plot $T_{al}$ and $T_f$ on the same graph.

**P10.2.** An airplane flying horizontally at 50 m/s and at an altitude of 300 m is to drop a food package weighing 2000 N to a group of people stranded in an inaccessible area resulting from an earthquake. A drag force, $\vec{D}$, acts on the package in the direction of the free stream, $\vec{V}$, as seen from the package (see Figure P10.1). We wish to determine the position of the package as a function of time and when the package hits the ground. Take

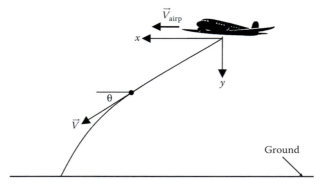

**FIGURE P10.1**
Path of a food package dropped from an airplane.

# Numerical Integration of Ordinary Differential Equations

$(x, y)$ to be the position of the package at time $t$ as seen from the position of the airplane at the time of release and

$(V_x, V_y)$ to be the horizontal and vertical components of the package velocity respectively.

Governing equations:

$$\frac{dV_x}{dt} = -\frac{C_d \rho V^2 A}{2M} \cos \vartheta \quad \text{(P10.2a)}$$

$$\frac{dV_y}{dt} = g - \frac{C_d \rho V^2 A}{2M} \sin \vartheta \quad \text{(P10.2b)}$$

$$\frac{dx}{dt} = V_x \quad \text{(P10.2c)}$$

$$\frac{dy}{dt} = V_y \quad \text{(P10.2d)}$$

$$\cos \vartheta = \frac{V_x}{V}, \text{ and } \sin \vartheta = \frac{V_y}{V} \quad \text{(P10.2e)}$$

$$V = \sqrt{V_x^2 + V_y^2} \quad \text{(P10.2f)}$$

where:
$C_d$ is the drag coefficient
$\rho$ is the air density
$M$ is the mass of package
$A$ is the frontal area of package

Initial conditions:

$$x(0) = 0, \ y(0) = 0, \ V_x(0) = 50 \text{ m/s}, \ V_y(0) = 0.$$

Use the following parameters:

$$C_d = 0.8, \ \rho = 1.225 \text{ kg/m}^3, \ A = 1.0 \text{ m}^2$$

Use MATLAB's ode45 function to solve for $(t, x, y, V_x, V_y)$ at intervals of 0.10 seconds for $0 \le t \le 10.0$ seconds.

1. Create plots of $x$ and $y$ versus $t$ both on the same graph.
2. Create plots of $V_x$ and $V_y$ versus $t$ both on the same graph.

3. Create a table containing $(t, x, y, V_x, V_y)$ at intervals of 0.10 seconds. Stop printing table the first time $y > 300$ m.
4. Use MATLAB's function interp1 to interpolate for the $(t, x, V_x, V_y)$ values when the package hits the ground. Print out these values.

**P10.3.** Figure P10.2 shows a third-order RLCC circuit. In order to run a time-domain transient analysis, we transform the circuit into three first-order differential equations which are

$$\frac{dv_{C1}}{dt} = \frac{1}{RC_1}(-v_{C1} + v_{C2} + Ri_L + v_S) \quad \text{(P10.3a)}$$

$$\frac{dv_{C2}}{dt} = \frac{1}{RC_2}(v_{C1} - v_{C2} - v_S) \quad \text{(P10.3b)}$$

$$\frac{di_L}{dt} = \frac{1}{L}(-v_{C1} + v_S) \quad \text{(P10.3c)}$$

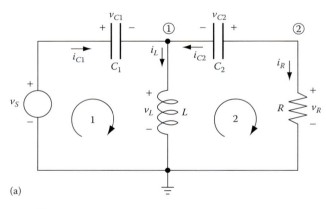

(a)

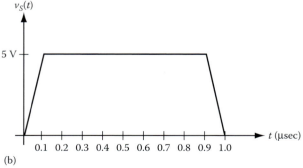

(b)

**FIGURE P10.2**
(a) A third-order RLCC circuit configuration and (b) pulse input.

We have chosen the three voltages and currents with derivative terms ($v_{C_1}$, $v_{C_2}$, and $i_L$) as the *state variables* for this problem.

Construct a MATLAB program using MATLAB's ode45 function to solve for the variables $v_{C_1}, v_{C_2}$, and $i_L$. Take $C_1 = 1$ μF, $C_2 = 0.001$ μF, $R = 100$ kΩ, $L = 0.01$ mH. Use a time interval of $0 \le t \le 5$ μs and a step size of 0.01 μs. Assume $v_S(t)$ is a 5V pulse starting at time $t = 0$ with rise time of 0.1 μs, an *on* time of 0.8 μs, and fall time of 0.1 μs (as shown in Figure P10.2). Initial conditions: $v_{C_1}(0) = 0$, $v_{C_2}(0) = 0$, and $i_L(0) = 0$.

Plot on separate graphs: $v_{C_1}(t)$, $v_{C_2}(t)$, $i_L(t)$, and $v_S(t)$ versus time.

**P10.4.** A small rocket with an initial mass of 350 kg, including a mass of 100 kg of fuel, is fired from a rocket launcher (see Figure P10.3). The rocket leaves the launcher at velocity $V_o$ and at an angle of $\theta_o$ with the horizontal. Neglect the fuel consumed inside the rocket launcher. The rocket burns fuel at the rate of 10 kg/s, and develops a thrust $T = 6000$ N. The thrust acts axially along the rocket and lasts for 10 s. Assume that the drag force also acts axially and is proportional to the square of the rocket velocity. The governing differential equations describing the position and velocity components of the rocket are as follows:

$$\frac{dV_x}{dt} = \frac{V_x T}{m\sqrt{V_x^2 + V_y^2}} - \frac{V_x K \sqrt{V_x^2 + V_y^2}}{m} \quad \text{(P10.4a)}$$

$$\frac{dV_y}{dt} = \frac{V_y T}{m\sqrt{V_x^2 + V_y^2}} - \frac{V_y K \sqrt{V_x^2 + V_y^2}}{m} - g \quad \text{(P10.4b)}$$

$$\frac{dx}{dt} = V_x \quad \text{(P10.4c)}$$

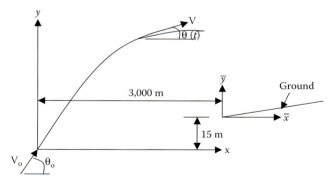

**FIGURE P10.3**
Rocket trajectory.

$$\frac{dy}{dt} = V_y \qquad (P10.4d)$$

$$V^2 = V_x^2 + V_y^2 \qquad (P10.4e)$$

where:
  $m$ is the mass of the rocket (varies with time)
  $V_x$, $V_y$ are the $x$ and $y$ components of the rocket's velocity relative to the ground
  $K$ is the drag coefficient
  $g$ is the gravitational constant
  $(x, y)$ are the position of the rocket relative to the ground
  $t$ is the time of rocket flight

The target lies on ground, which has a slope of 5%. The ground elevation relative to the origin of the coordinate system of the rocket is given by

$$y_g = 15 + 0.05(x - 3000) \qquad (P10.4f)$$

Using Equations P10.4a through P10.4d, write a computer program in MATLAB using the MATLAB's ode45 function that solves for $x$, $y$, $V_x$, and $V_y$ for $0 \le t \le 60$ seconds in steps of 0.01 s. Use Equation P10.4f to solve for $y_g$. Take $x(0) = 0$, $y(0) = 0$, $V_x(0) = V_o\cos\theta_o$, $V_y(0) = V_o\sin\theta_o$, $V_o = 150$ m/s, $\theta_o = 60°$, $K = 0.045$ N–s²/m², and $g = 9.81$ m/s².

1. Print out a table for $x$, $y$, $y_g$, $V_x$, $V_y$ every 1.0 second.
2. Use MATLAB to plot $x$, $y$, and $y_g$ versus $t$ on the same graph and $V_x$, $V_y$, versus $t$ on the same graph.
3. Assume a linear trajectory between the closest two data points where the rocket hits the ground. The intersection of the two straight lines gives the $(x, y)$ position of where the rocket hits the ground.

**P10.5.** We wish to examine the time temperature variation of a fluid, $T_f$, enclosed in a container with a heating element and a thermostat. The walls of the container are pure copper. The fluid is engine oil, which has a temperature $T_f$ that varies with time. The thermostat is set to cut-off power from the heating element when the $T_f$ reaches 65°C and to resume supplying power when $T_f$ reaches 55°C. The outside room temperature, $T_\infty$, remains constant at 15°C.
Wall properties:

$$k = 386.0 \text{ w/m-C}, \quad c = 0.3831 \text{ kJ/kg-C}, \quad \rho = 8954 \text{ kg/m}^3$$

Engine oil properties:

$$k = 0.137 \text{ w/m-C}, \quad c = 2.219 \text{ kJ/kg-C}, \quad \rho = 840 \text{ kg/m}^3$$

The inside size of the container is (0.5 m × 0.5 m × 0.5 m)

# Numerical Integration of Ordinary Differential Equations

The wall thickness is 0.01 m. Thus, the

Inside surface area, $A_{s,i} = 1.5$ m²
Outside surface area, $A_{s,o} = 1.5606$ m²
Engine oil volume, $V_{oil} = 0.125$ m³
Wall volume, $V_{wall} = 0.0153$ m³

The power, $Q$, of the heating element $= \begin{cases} 20000 \text{ W when } t_f < 55°C \\ 0 \text{ when } t_f > 65°C \end{cases}$

The inside convective heat transfer coefficient, $h_i = 560$ W/m²-C
The outside convective heat transfer coefficient, $h_o = 110$ W/m²-C

Using a lump parameter analysis (assuming that the engine oil is well mixed) in heat transfer, the governing equations describing the time temperature variation of both materials are as follows:

$$\frac{d\theta_f}{dt} = -a_1(\theta_f - \theta_w) + a_5 \tag{P10.5a}$$

$$\frac{d\theta_w}{dt} = a_2(\theta_f - \theta_w) - a_3\theta_w = a_2\theta_f - (a_2 + a_3)\theta_w \tag{P10.5b}$$

where:

$$\theta_f = T_f - T_\infty \tag{P10.5c}$$

$$\theta_w = T_w - T_\infty \tag{P10.5d}$$

$$a_1 = \frac{h_i A_{s,i}}{m_f c_f}, \quad a_2 = \frac{h_i A_{s,i}}{m_w c_w}, \quad a_3 = \frac{h_o A_{s,o}}{m_w c_w}, \quad a_4 = a_2 + a_3, \quad a_5 = \frac{Q}{m_f c_f} \tag{P10.5e}$$

Initial conditions:

$$T_f(0) = Tw(0) = 15°C$$

$$T_\infty = 15°C$$

Using MATLAB's ode45 function to construct a simulation of this system. Run the time for 3600 seconds. Print out values of $T_f$ and $T_w$ versus $t$ at every 100 seconds. Construct plots of $T_f$ and $T_w$ versus $t$. Use tspan = 0:1:3600.

**P10.6.** We wish to determine the altitude and velocity of a helium filled spherically shaped balloon as it lifts off from its mooring. We will assume that atmospheric conditions can be described by the U.S. Standard Atmosphere. We will assume that there is no change in the balloon's volume. For a complete derivation of the equations given below, see Project P7.9 in Reference 1. The governing equations describing the motion of the balloon are

$$\frac{dz}{dt} = V \qquad \text{(P10.6a)}$$

$$\frac{dV}{dt} = \frac{1}{M}(B - W - \text{sgn} * D) \qquad \text{(P10.6b)}$$

where:
- $z$ is the altitude of the centroid of the balloon
- $V$ is the vertical velocity of the balloon
- $t$ is the time
- $B$ is the buoyancy force acting on the balloon (varies with altitude)
- $M$ is the total mass of the balloon material, ballast, and the gas
- $W$ is the total weight of the balloon material, ballast, and the gas $= Mg$
- $D$ is the drag on the balloon
- sgn $= +1$, if $(dz/dt) \geq 0$ and sgn $= -1$, if $(dz/dt) < 0$

The U.S. Standard atmosphere as applied to this balloon problem consists of the following governing equations:

$$\frac{dp}{dt} = -\frac{p}{RT} gV \qquad \text{(P10.6c)}$$

$$T = T_i - \lambda z \qquad \text{(P10.6d)}$$

$$\rho = \frac{p}{RT} \qquad \text{(P10.6e)}$$

where:
- $p$ is the outside air pressure at the centroid of the balloon
- $\rho$ is the outside air density at the centroid of the balloon
- $g$ is the gravitational constant that varies with altitude
- $R$ is the gas constant for air
- $T$ is the outside air temperature at the centroid of the balloon
- $T_i$ is the temperature at the earth's surface $= 288.15$ (K).
- $\lambda$ is the lapse rate

The buoyancy force, $B$, is given by

$$B = \rho g \upsilon \qquad \text{(P10.6f)}$$

and

$$g = g_0 \left( \frac{r_e}{z + r_e} \right) \qquad \text{(P10.6g)}$$

where:
- $\upsilon$ is the volume of the balloon $= (4/3) r_b^3$
- $r_b$ is the radius of the balloon

# Numerical Integration of Ordinary Differential Equations

$r_e$ is the radius of the earth
$g_0$ is the gravitational constant near the earth's surface
$g$ is the gravitational constant at an elevation of the centroid of the balloon

For low Reynolds Number, Re, less than 0.1, the drag force, $D$, is given by Stokes formula, which is

$$D = 6\pi\mu V r_b \tag{P10.6h}$$

For flow speeds with Re > 0.1, use

$$D = C_d \frac{\rho}{2} V^2 A \tag{P10.6i}$$

where:
$C_d$ is the drag coefficient
$A$ is the frontal area of the balloon = $\pi r_b^2$

The drag coefficient, $C_d$, is given by

$$C_d = \frac{24}{\text{Re}} + \frac{6}{1.0 + \sqrt{\text{Re}}} + 0.4 \tag{P10.6j}$$

where

$$\text{Re} = \frac{2\rho V r_b}{\mu} \tag{P10.6k}$$

and $\mu$ is the fluid viscosity

The fluid viscosity, $\mu$, can be determined by the Sutherland formula, which is

$$\mu = \mu_0 \left(\frac{T}{T_0}\right)^{1.5} \left(\frac{T_0 + S}{T + S}\right) \tag{P10.6l}$$

For air, $S$ = 110.4 K, $\mu_0$ = 1.71e-5 N-s/m², $T_0$ = 273 K.

Write a computer program, using MATLAB's `ode45` function that will determine the balloon's altitude as a function of time.

**NOTE:** Equations P10.6a through P10.6c represent a system of three first-order ordinary differential equations that can be solved by MATLAB's `ode45` function.

Create plots of $z$ versus $t$, $v$ versus $T$, and $p$ versus $t$. Use the following values:

$M$ = 2200 kg, $r_b$ = 7.816 m, $\upsilon$ = $4/3\pi r_b^3$ m³, $R$ = 287 J/(kg-K), $T_i$ = 288.15 K,

$\lambda$ = 0.0065 (K/m), $g_0$ = 9.81 m/s², and $r_e$ = 6371e+3 m

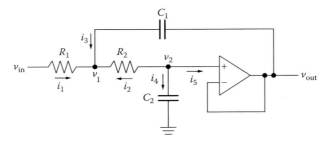

**FIGURE P10.4**
Sallen–key circuit.

Use a tspan = 0.0:0.1:1000 and the following initial conditions:

$$z(0) = r_b,\ V(0) = 0,\ p(0) = 1.0132e+5$$

**P10.7.** The Sallen–Key circuit (Figure P10.4) is commonly used to implement second-order (or higher) filters. The following equations model the circuit using ordinary differential equations. It is assumed that the op amp is ideal resulting in $v_2 = v_{out}$ and $i_5 = 0$. Applying Kirchhoff's Current Law at the nodes labeled $v_1$ and $v_2$, and the constituent relations for resistors ($v_R = i_R R$) and capacitors ($i_C = C(dv_C/dt)$), the following equations are obtained:

$$\frac{dv_{out}}{dt} = \left(\frac{-1}{R_2 C_2}\right) v_{out} + \left(\frac{1}{R_2 C_2}\right) v_1 \quad \text{(P10.7a)}$$

$$\frac{dv_1}{dt} = \left(\frac{1}{R_2 C_1} - \frac{1}{R_2 C_2}\right) v_{out} + \left(\frac{1}{R_2 C_2} - \frac{1}{R_1 C_1} - \frac{1}{R_2 C_1}\right) v_1 + \left(\frac{1}{R_1 C_1}\right) v_{in} \quad \text{(P10.7b)}$$

1. Solve for $v_{out}$ and $v_1$ using MATLAB's ode45 function. Assume that the input to the circuit $v_{in}$ is a step voltage that changes from 0 V to 1 V at time $t = 0^+$. Assume the following values for the circuit elements: $R_1 = 5000\ \Omega$, $R_2 = 5000\ \Omega$, $C_1 = 2200$ pF, $C_2 = 1100$ pF. Use a time interval of $t = [0,\ 100\ \mu s]$ s and assume $v_{out}(0) = v_1(0) = 0$.

2. Find the impulse response of the circuit by first creating a MATLAB function pulse(t) that returns the following values:

$$\text{pulse(t)} = \begin{cases} 10^6 & \text{for } 0 < t < 10^{-6} \\ 0 & \text{otherwise} \end{cases}$$

Then, solve for $v_{out}$ and $v_1$ using MATLAB's ode45 function where $v_i = $ pulse(t). Use the same component values, time interval, and initial conditions as in part 1.

3. Plot the step response (from part 1) and the impulse response (from part 2) on the same set of axes. What relationship can you see between the two?

# Numerical Integration of Ordinary Differential Equations

**P10.8.** Exercise E2.4 involved a basketball player shooting a basketball toward the hoop. The basketball was released 6 m from the center of the hoop with a velocity, $V_o$, and making an angle of 40° with the horizontal (see Figure 2.20). Equations describing the motion of the basketball based on Newton's second law and neglecting drag were given. The solution obtained in Exercise E2.4 was that if $V_o = 8.7098$ m/s the basketball would reach the center of the hoop in 0.8993 s. We now want to include drag in determining the motion of the basketball. We have assumed that the drag is in the opposite direction of the ball's motion. The governing equations become

$$\frac{dV_x}{dt} = -\frac{\rho A C_d V_x \sqrt{V_x^2 + V_y^2}}{2m} \tag{P10.8a}$$

$$\frac{dV_y}{dt} = -\text{sgn} \times \frac{\rho A C_d V_y \sqrt{V_x^2 + V_y^2}}{2m} - g \tag{P10.8b}$$

$$\frac{dx}{dt} = V_x \tag{P10.8c}$$

$$\frac{dy}{dt} = V_y \tag{P10.8d}$$

$$V = \sqrt{V_x^2 + V_y^2} \tag{P10.8e}$$

where:
$\rho$ is the density of the air
A is the frontal area of the basketball
$C_d$ is the drag coefficient
m is he mass of the basketball
g is the gravitational constant
($V_x$, $V_y$) are the x and y components of the velocity and (x, y) are the horizontal and vertical positions of the basketball
sgn = −1.0 if $V_y$ < 0 and sgn = 1.0 if $V_y$ > 0

Using MATLAB's ode45 function, determine the x-position of the basketball when it reaches the height of the hoop. Use the following parameters:

$m = 0.623445$ kg, $\rho = 1.225$ kg/m$^3$, $A = 0.04476$ m$^2$, $C_d = 0.25$, $g = 9.81$ m/s$^2$,

$0 \le t \le 1$ s in steps of 0.01 s, and the coordinates of the center of the hoop

$(x_h, y_h) = (6, 3.048)$ m, $y(0) = 1.98$ m and $x(0) = 0$.

Hint: MATLAB's ode45 function should give you $x(i)$, $y(i)$, $V_x(i)$, $V_y(i)$, for $i = 1$:length($t$). Determine the first $i$ value when $y(i) < 3.048$ and $V_y(i) < 0$. Use that value of $i$ and the one before it to interpolate the x-position when $y = 3.048$. Would the basketball hit the rim of the hoop? The rim radius is 0.2286 m.

# Reference

1. Bober, W., *Introduction to Numerical and Analytical Methods with MATLAB for Engineers and Scientists*, CRC Press, Boca Raton, FL, 2014.

# 11

# Boundary Value Problems of Ordinary Differential Equations

## 11.1 Introduction

When an ordinary differential equation involves boundary conditions instead of initial conditions, then a numerical approach is most often used to solve the problem. In a boundary value problem, we essentially need to *fit* a solution into the known boundary conditions as opposed to simply integrating from the initial conditions. An example of this type of problem is the temperature of a bar subjected to known different temperatures at the ends as it looses heat along the bar by natural convection. Another example would be the deflection of a beam due to an applied load along the beam and where the boundary conditions at both ends of the beam are specified. Another example of this type of problem is the determination of the electric field between the plates of a capacitor with a known charge density between the plates and a fixed voltage across the plates. In these three examples, a solution is found by numerically solving a second-order, nonhomogeneous ordinary differential equation using *finite difference methods*.

## 11.2 Difference Formulas

To numerically solve a boundary value problem involving an ordinary, linear, and differential equation, we will need the difference formulas obtained by Taylor series expansion using just a few terms. The finite difference method first involves subdividing the independent variable domain into $N$ subdivisions. The finite difference formulas that will be used are tabulated in Table 11.1.

In the following table, $y_i$ is the value of $y$ at position $x_i$ and $\Delta x = x_{i+1} - x_i$.

## TABLE 11.1
### Summary of Finite Difference Formulas for Boundary Value Problems

| | |
|---|---|
| $y_i' = \dfrac{y_{i+1} - y_i}{\Delta x}$ | First-order forward difference formula. Usually used for a $y'$ boundary condition at the beginning of domain. |
| $y_i' = \dfrac{y_i - y_{i-1}}{\Delta x}$ | First-order backward difference formula. Usually used for a $y'$ boundary condition at end of the domain. |
| $y_i'' = \dfrac{y_{i+1} + y_{i-1} - 2y_i}{\Delta x^2}$ | Second-order central difference formula for the second derivative in the interior of the domain. |

### Example 11.1

In this example we consider a bar having a circular cross section that is subjected to known temperatures at the two ends and which looses heat along the bar by natural convection to its surroundings (see Figure 11.1). The governing equations describing the temperature along the bar is given by the following formula:

$$\frac{d^2T}{dx^2} = \frac{hP}{Ak}(T - T_\infty) \qquad (11.1)$$

where:
 $T$ is the temperature of the bar at position $x$
 $h$ is the convective heat transfer coefficient
 $k$ is the thermal conductivity of the bar material
 $P$ is the bar perimeter
 $A$ is the bar cross-sectional area
 $T_\infty$ is the temperature of the surrounding air

Applying the finite difference formulas to the problem gives the following set of equations:

$$\frac{T_{i+1} + T_{i-1} - 2T_i}{\Delta x^2} = \frac{hP}{kA}(T_i - T_\infty) \qquad (11.2)$$

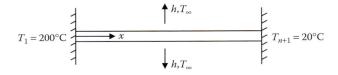

**FIGURE 11.1**
Bar subjected to different end temperatures and losing heat to the surroundings.

# Boundary Value Problems of Ordinary Differential Equations 229

We will assume that the $x$ domain is divided into $N$ subdivisions and that $T_1 = 200°C$ and $T_{N+1} = 20°C$. Then the set of equations become

$$T_1 = 200 \qquad (1)$$

$$T_1 + T_3 - 2T_2 - \frac{hP\Delta x^2}{kA}T_2 = \frac{hP\Delta x^2}{kA}T_\infty \qquad (2)$$

$$T_2 + T_4 - 2T_3 - \frac{hP\Delta x^2}{kA}T_3 = \frac{hP\Delta x^2}{kA}T_\infty \qquad (3)$$

$$T_3 + T_5 - 2T_4 - \frac{hP\Delta x^2}{kA}T_4 = \frac{hP\Delta x^2}{kA}T_\infty \qquad (4)$$

.

.

$$T_{N-1} + T_{N+1} - 2T_N - \frac{hP\Delta x^2}{kA}T_N = \frac{hP\Delta x^2}{kA}T_\infty \qquad (N)$$

$$T_{N+1} = 20 \qquad (N+1)$$

The above set of algebraic, linear equations can be solved by using MATLAB®'s `inv` function or MATLAB's Gauss-Elimination function. In Example 7.3, we discussed a systematic method for solving this type of problem. The set of equations can be expressed as the matrix equation $\mathbf{AT} = \mathbf{C}$, where

$$\mathbf{T} = \begin{bmatrix} T_1 \\ T_2 \\ \vdots \\ T_n \end{bmatrix}, \quad \mathbf{A} = \begin{bmatrix} a_{1,1} & a_{1,2} & \cdots & a_{1,n} \\ a_{2,1} & a_{2,2} & \cdots & a_{2,n} \\ \vdots & \vdots & \ddots & \vdots \\ a_{n,1} & a_{n,2} & \cdots & a_{n,n} \end{bmatrix}, \quad \mathbf{C} = \begin{bmatrix} c_1 \\ c_2 \\ \vdots \\ c_n \end{bmatrix} \qquad (11.3)$$

In the above set of equations, the coefficient matrix $\mathbf{A}$ is made up of elements $a_{i,j}$ where the first index is the equation number and the second index is the same as the index of the unknown temperature, $T_j$ that the $a_{i,j}$ element is associated with. For example, in Equation 2,

$$a_{2,1} = 1,\ a_{2,3} = 1 \text{ and } a_{2,2} = -\left(2 + \frac{hP\Delta x^2}{kA}\right).$$

In this expression, $A$ is the cross-sectional area of the bar and not the coefficient matrix, $\mathbf{A}$. From the pattern of the equation set, we can assign all the $a_{i,j}$ terms within one `for` loop. This is done in the following program:

We will use the following parameters:

$$k = 386 \text{ W/m-C}, h = 60 \text{ W/m}^2\text{-C}, L = 0.5 \text{ m}, N = 50, D = 0.2 \text{ cm}$$

The following program creates a table and plot of $T$ versus $x$.

```
% Example_11_1.m
% This program determines the temperature in a bar having
% different end temperatures and subjected to convective
% heat transfer.
% Units for k are W/m-C, units for h are W/m^2-C
% units for T are C, units for L are m, units for D are cm.
clear; clc;
k=386; h=60; L=0.5; D=0.2e-2;
P=pi*D; A=pi/4*D^2; N=50; dx=L/N;
T(1)=200; T(N+1)=20; Tinf=20;
C1=h/k*P/A*dx^2;
a(1,1)=1; C(1)=200;
a(N+1,N+1)=1; C(N+1)=Tinf;
x=0:dx:L;
for i=2:N
    a(i,i-1)=1; a(i,i+1)=1; a(i,i)=-(2+C1); C(i)=C1*Tinf;
end
T=inv(a)*C';
plot(x,T), xlabel('x(m)'), ylabel('T(C)'), grid, title('T vs. x');
fprintf('x(m)       T(C)    \n')
fprintf('--------------------\n');
for i=1:2:N+1
    fprintf('%4.2f       %7.2f \n',x(i),T(i));
end
```

**Program Results:**

```
x(m)       T(C)
--------------------
0.00       200.00
0.02       185.00
0.04       171.03
0.06       158.02
0.08       145.91
0.10       134.64
  .          .
  .          .
0.40       35.05
0.42       31.55
0.44       28.31
0.46       25.32
0.48       22.55
0.50       20.00
>>
```

See Figure 11.2.

# Boundary Value Problems of Ordinary Differential Equations

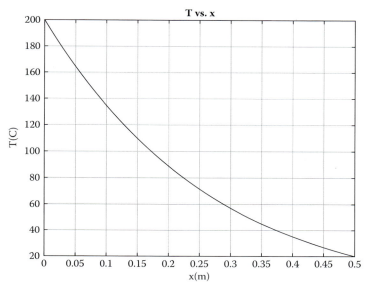

**FIGURE 11.2**
Plot of temperature, T versus position x.

## Exercises

**E11.1.** Repeat Example 11.1, but this time replace the boundary condition at $x = L$, with

$$-k\frac{dT}{dx}(L) = h(T(L) - T_\infty) \text{ giving}$$

$$k\frac{T_{N+1} - T_N}{\Delta x} + hT_{N+1} = hT_\infty$$

**E11.2.** In this exercise, we consider the deflection of a beam subjected to a uniform load, of weight, $w/m$ (for more details on this subject, see Example 8.1 in Reference 1. Consider the beam shown in Figure 11.3. The governing equation for the deflection of a beam is

$$\frac{d^2y}{dx^2} = \frac{M(x)}{EI(x)} \tag{11.4}$$

where:
  $y$ is the deflection of beam
  $M$ is the internal bending moment
  $E$ is the modulus of elasticity of beam material
  $I$ is the moment of inertia of an area

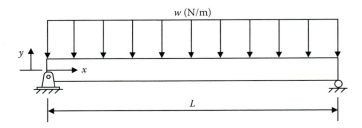

**FIGURE 11.3**
Deflection of a uniformly loaded beam.

To obtain the finite difference form of the governing equation, subdivide the $x$-axis into $N$ subdivisions, giving $x_1, x_2, x_3, \ldots, x_{N+1}$.

Let the deflections at these points be: $y_1, y_2, y_3, \ldots, y_{N+1}$.

The finite difference formula for $d^2y/dx^2$, as shown in Table 11.1 is

$$\frac{d^2y}{dt^2}(x_n) = \frac{y_{n+1} + y_{n-1} - 2y_n}{\Delta x^2} \tag{11.5}$$

Thus, the governing differential equation becomes

$$\frac{y_{n+1} + y_{n-1} - 2y_n}{\Delta x^2} = \frac{M_n}{EI}$$

or

$$\frac{1}{2} y_{n-1} + \frac{1}{2} y_{n+1} - y_n = \frac{M_n \Delta x^2}{2EI}, \text{ for } n = 2,3,4,\ldots, N \tag{11.6}$$

The boundary conditions are

$$y_1 = 0 \tag{11.7}$$

$$y_{N+1} = 0 \tag{11.8}$$

The bending moment $M_n$ for this problem is

$$M_n = \frac{wL}{2} x_n - \frac{w x_n^2}{2} \tag{11.9}$$

Determine the deflection, $y_i$ for $i = 1:N + 1$. Create a plot $y$ versus $x$ and print a table consisting of $y_i$ and $x_i$. Also print out the obtained maximum deflection. Use the following parameters:

$$L = 3 \text{ m}, w = 40 \text{ kN/m}, EI = 1.5 \times 10^3 \text{ kN-m}^2, N = 30$$

# Boundary Value Problems of Ordinary Differential Equations

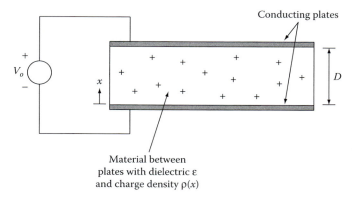

**FIGURE 11.4**
Parallel plate capacitor with constant applied voltage.

**E11.3.** Figure 11.4 shows a parallel plate capacitor with constant applied voltage $v_o$ and a fixed charge density $\rho$ between the plates. For cases with planar symmetry such as the parallel plate capacitor where the charge density only changes in the $x$-direction (i.e., there is no $y$ or $z$ dependency), then Poisson's equation describing the electric potential $\Phi$ reduces to an ordinary differential equation:

$$\frac{d^2\Phi(x)}{dx^2} = -\frac{\rho(x)}{\varepsilon} \tag{11.10}$$

where:
$\Phi(x)$ is the electric potential (in volts)
$\rho(x)$ is the $x$-dependent charge density (in coul/m³)
$\varepsilon$ is the dielectric constant for the material between the plates

We wish to solve for $\Phi(x)$ between the plates of the capacitor shown in Figure 11.4 with a plate separation of $D$ meters, $\rho(x) = \rho_o(x-D)^2$ and boundary conditions $\Phi(0) = 0$ and $\Phi(D) = v_o$.
Substituting the expression for $\rho(x)$ into Equation 11.10, we obtain

$$\frac{d^2\Phi}{dx^2} = -\frac{\rho_o}{\varepsilon}(x-D)^2 = -\frac{\rho_o}{\varepsilon}(x^2 - 2Dx + D^2) \tag{11.11}$$

Equation 11.11 can readily be solved analytically and the solution [2] is

$$\Phi(x) = -\frac{\rho_o}{\varepsilon}\left[\frac{1}{12}x^4 - \frac{D}{3}x^3 + \frac{D^2}{2}x^2 - \left(\frac{\varepsilon v_o}{D\rho_o} + \frac{D^3}{4}\right)x\right] \tag{11.12}$$

Solve Equation 11.11 numerically and compare the numerical solution with the exact solution. Create a plot of $\Phi$ versus $x$ and a table consisting of $x$, $\Phi$

obtained numerically, and Φ obtained by Equation 11.12. Use the following parameters:

$$N = 40, D = .0004 \text{ m}, rho = 1\times 10^4 \text{coulomb/m}^3,$$
$$epsilon = 1.04\text{e-}12, v_o = 5V$$

**Projects**

**P11.1.** For the beam shown in Figure P11.1, determine the beam deflection, $y(x)$, by the finite difference method utilizing MATLAB's inverse matrix function or the MATLAB's Gauss-Elimination Method. Print the results in a table format. Also determine the approximate maximum deflection. Use the following parameters:

$$P = 10 \text{ kN}, EI = 1.5\text{e+}3 \text{ kN-m}^2, L = 10 \text{ m}, a = 6 \text{ m},$$
$$0 \le x \le 10.0 \text{ m in steps of } 0.1 \text{ m}$$

For this configuration:

$$M_i = \begin{cases} \dfrac{P(L-a)}{L} x_i & \text{for } 0 \le x_i \le a \\ \dfrac{P(L-a)}{L} x_i - P(x_i - a) & \text{for } a < x_i \le L \end{cases}$$

**P11.2.** For the beam shown in Figure P11.2, determine the beam deflection, $y(x)$, by the finite difference method utilizing MATLAB's inverse matrix function or the MATLAB's Gauss-Elimination method. Print the results in a

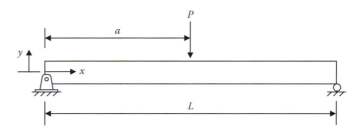

**FIGURE P11.1**
Deflection of a beam subjected to a concentrated load.

# Boundary Value Problems of Ordinary Differential Equations

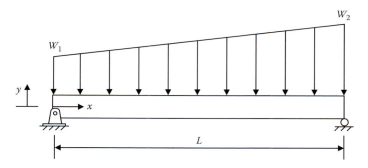

**FIGURE P11.2**
Deflection of a beam subjected to a linear increasing load.

table format. Also determine the approximate maximum deflection. Use the following parameters:

$w_1 = 10$ kN/m, $w_2 = 20$ kN/m, $EI = 1.5e+3$ kN-m², $L = 10$ m,
$0 \le x \le 10.0$ m in steps of 0.1 m.

Hint: The load can be considered as the sum of a uniform load and a triangular load. For the triangular load, the resultant force equals $(w_2-w_1)L/2$ located $2L/3$ from the apex of the triangle. This results in the following equation for $M(x)$:

$$M(x) = \left(\frac{w_1 L}{2} + (w_2 - w_1)\frac{L}{6}\right)x - \frac{w_1 x^2}{2} - (w_2 - w_1)\frac{x^3}{6L}$$

## References

1. Bober, W., *Introduction to Numerical and Analytical Methods with MATLAB for Engineers and Scientists*, CRC Press, Boca Raton, FL, 2014.
2. Bober, W., Stevens, A., *Numerical and Analytical Methods with MATLAB for Electrical Engineers*, CRC Press, Boca Raton, FL, 2012.

# Appendix: Greek Letters and Special Characters in MATLAB® Plots

MATLAB® allows the use of Greek and special characters in its plot headings and labels. The method for doing this is based on the TeX formatting language [1] and is summarized in this appendix.

MATLAB provides the functions `title`, `xlabel`, `ylabel`, and `text` for adding labels to plots. These labels can include Greek and special characters by applying the character sequences as shown in Table A.1. These sequences all begin with the backslash character (\) and can be embedded in any text string argument to `title`, `xlabel`, `ylabel`, and `text`. Subscripts and superscripts may also be applied by using the _ and ^ operators. For example, the sequence $V_o$ is written as V_o and $10^6$ is written as 10^6. If the subscripts or superscripts are multiple characters, then use curly braces to delimit the string to be subscripted, for example, $V_{out}$ is generated with V_{out}.

**Example A.1**

The following MATLAB script shows how to use special characters in a plot:

```
% Example_A_1.m
% This script shows example usage of special characters in MATLAB plots.
% Plot a 1MHz sine wave over the interval 0<t<2 microsec
t = 0:2e-8:2e-6;
fo = 1e6;
xout = sin(2*pi*fo*t);
plot(t*1e6,xout);
title('Plot of sin(2\pif_{o}t) for f_{o}=10^6 Hz');
xlabel('time (\museconds)');
ylabel('x_{out}(t)');
text(1.5,0.3,'\omega = 2\pi \times f_{o}');
```

The resulting plot is shown in Figure A.1.

**TABLE A.1**

**Greek Letters**

| Greek Letter | Greek Symbol |
| --- | --- |
| \alpha | $\alpha$ |
| \beta | $\beta$ |
| \gamma | $\gamma$ |
| \delta | $\delta$ |
| \epsilon | $\varepsilon$ |
| \zeta | $\zeta$ |
| \eta | $\eta$ |
| \theta | $\theta$ |
| \vartheta | $\vartheta$ |
| \lambda | $\lambda$ |
| \mu | $\mu$ |
| \nu | $\nu$ |
| \rho | $\rho$ |
| \sigma | $\sigma$ |
| \tau | $\tau$ |
| \phi | $\phi$ |
| \omega | $\omega$ |
| \Gamma | $\Gamma$ |
| \Delta | $\Delta$ |
| \Phi | $\Phi$ |
| \Omega | $\Omega$ |

# Appendix

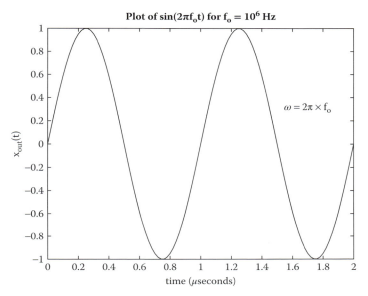

**FIGURE A.1**
Example usage of Greek letters, special characters, subscripts and superscripts in a MATLAB plot.

## Reference

1. Knuth, D.E., *The TeXbook*, Addison Wesley, New York, 1984.

# Review Answers

## Review 1.1

1. List several ways engineers use the computer.
    a. To solve mathematical models of physical phenomenon.
    b. Storing and reducing experimental data.
    c. Controlling machines.
    d. Communicating with other engineers on a particular project.
2. List several areas of interest for engineers.
    a. Designing new products.
    b. Improve performance of existing products.
    c. Improving manufacturing efficiency.
    d. Minimizing costs of production.
    e. Minimizing power consumption.
    f. Research on developing new products.
3. List several methods that can be used in the design of new products.
    a. Full-scale experiments.
    b. Small-scale model experiments.
    c. A mathematical model describing the phenomenon of interest.
4. Which method mentioned in item 3 is the least expensive?
   The mathematical model is the least expensive.
5. List several components of a typical desktop/laptop computer system.
    a. Input devices: keyboard, mouse, microphone.
    b. Central processing unit consisting of a control unit, an arithmetic logic unit and registers.
    c. Memory and storage unit consisting of main memory which is used for temporary storage of programs and data.
    d. Secondary memory consisting of a hard drive, an optical drive (a CD or a DVD), and a flash drive.
    e. Output devices: monitor, printer, speakers.
    f. Operating systems: Windows 10, MacOS, Linux.
6. Name several computer languages used today and in the past by Engineers.
   Fortran, C/C++, MATLAB, Pascal.

7. What is the lowest level computer language and what numbering system does it use?

Machine language. It uses the binary system of numbers.

8. For engineers what is the principle advantage of MATLAB® over some of the other computer programming platforms?

MATLAB has built-in functions that solve many different types of mathematical problems that other computer platforms do not have.

9. List several items that are recommended in developing a computer program.
   a. List the algebraic equations involved in the project.
   b. Create a flow chart or a program outline.
   c. Write the program using the list of algebraic equations and the program outline or flow chart.
   d. Run the program and correct any syntax errors.

10. List several items that can be considered building blocks available in developing a computer program in MATLAB.
    a. Assignments (Arithmetic statements).
    b. Input/output statements.
    c. Loop statements (`for` loop and `while` loop).
    d. Conditional operatives/alternative path statements (`if-else`, `if-elseif-else` statements).
    e. Functions (built-in and self-written).

## Review 2.1

1. What are the two alternative ways to start the MATLAB program?

   If available, start the MATLAB program by double-clicking on the MATLAB icon on the Window's desktop. If not available,
   a. For Windows versions earlier than Windows 10, go to the Window's *Start* menu, click on *All Programs*, find the *MATLAB* program among the list of available programs and click on it. This will open up the MATLAB desktop.
   b. For Windows version 10, click on the Windows icon on the left bottom of the screen and search for the MATLAB program and click on it. This will open the MATLAB desktop.

2. What are the windows in the MATLAB's default desktop?

   The main windows are the Command Window, Current Folder, and Workspace.

# Review Answers

3. It is best to write a MATLAB script (program) in the Editor Window. From MATLAB's default desktop how does one open the script window?

   To Open the Editor Window, click on the *New Script icon* in the Toolstrip in MATLAB's desktop.

4. After you have completed writing a script in the appropriate window, what is the next step?

   Click on the save icon in the Toolstrip. In the window that opens, select the folder in which the script is to be saved and in the *File Name* box type in the name of the script with the *.m* extension.

5. Name two ways to execute a script.
   a. In the Editor Window, click on arrow located just above the *Run icon* in the Toolstrip. In the Editor Window the arrow is green.
   b. In the Command Window after the MATLAB prompt (>>) sign, type in the script name (without the *.m* extension).

6. What happens if you attempted to execute a script and the script is not in the folder listed in the current folder Toolbar?

   A dialog box will open giving you the choice of changing the folder listed in the path box or adding the folder containing your script to the MATLAB path.

7. In MATLAB, what is the file name extension for saved scripts?

   The file name extension is *.m*.

8. How does one establish a comment line in a script?

   By placing a % sign in front of a statement makes it a comment line.

## Review 2.2

1. List at least two conditions in selecting a name for a variable.
   a. Variable name must start with a letter.
   b. Variable names can only contain letters, numbers, and the underscore character.

2. Finish the following statement. An arithmetic statement may involve … constants, variables, arithmetic operators, and elementary MATLAB functions and self-written functions.

3. What can be said about the variables that appear on the right side of an arithmetic statement?

   All variables on the right side of an arithmetic statement must be previously defined (given a value) in the program.

4. List the arithmetic operators in MATLAB.

   The arithmetic operators in MATLAB are

   + addition
   − subtraction
   * multiplication
   / division
   ^ exponentiation

5. What is the order in which an arithmetic statement will be carried out?

   First all operations enclosed within parentheses will be carried out in the following order: exponentiation, multiplication, and division, then addition and subtraction. Then proceeding from left to right, the operations will be carried out in the same order as listed above.

6. What is MATLAB's command for

   a. π.                                                              pi
   b. e.                                                              exp
   c. ln.                                                             log
   d. sine function in radians.                                       sin()
   e. sine function in degrees.                                       sind()
   f. $sin^{-1}$ function.                                            asin()
   g. The number of elements in a vector.                             lenth()
   h. The size of a matrix (the number of rows and columns).          size()
   i. The sum of the elements in a vector.                            sum()
   j. The maximum element in a vector.                                max()
   k. Preallocating the size of a 3 × 3 matrix.                       zeros(3)

7. What is the purpose of placing a semicolon at the end of a variable assignment?

   Placing a semicolon at the end of a variable assignment avoids the variable assignment to be echoed to the screen.

8. What is the command that will clear the Command Window?

   The command `clc` clears the Command Window.

9. What is the basic data structure in MATLAB?

   The basic data structure in MATLAB is a matrix.

10. Name two functions of the colon operator.

    a. The colon operator may be used to create a new matrix from an existing matrix.
    b. The colon operator can also be used to generate a series of numbers.

# Review Answers 245

## Review 2.3

1. Name two commands that will result in printing to the screen.
   a. `fprintf()`
   b. `display()`
2. What is the command that will move the cursor to the next line?
   `\n`.
3. What is the format that will print a floating point variable to 10 spaces and to three decimal points?
   `% 10.3f`
4. What is the format that will print a floating point variable in scientific notation to 12 spaces and to four decimal points?
   `% 12.4e`
5. What are the commands necessary to print to a file?
   a. `fo=fopen('file_name','w');`
   b. `fprintf(fo,'format \n',variables);`
6. What is the command to create a linear plot of $y$ versus $x$ and what type of variable must $x$ and $y$ be?
   The command is `plot(x,y);`
   Variables $x$ and $y$ must be vectors.
7. What are the commands that will label the $x$- and $y$-axis and provide a title to a plot?
   `xlabel('x'), ylabel('y'), title('y vs. x')`

## Review 2.4

1. What is the objective in using a `for` loop?
   The objective of a `for` loop is to repeat a series of statements with just a few lines of code.
2. What is the syntax of a `for` loop?
   `for index variable = starting value: step size: final value`
3. Should table headings that are not to be repeated be inside a `for` loop?
   No.

4. If the index of a `for` loop is used to select an element of a vector or a matrix, what variable type should the `for` loop index be?

   It should be an integer.

5. What other statement type can be used to create a loop?

   The `while` loop.

6. What is the major difference between a `for` loop and a `while` loop?

   The syntax of the `for` loop generates its own index. If a program requires an index, the program containing the `while` loop must generate an index.

## Review 2.5

1. Name four commands that can be used in a script to input data into the workspace. Also list where the data are located.

   The commands that can be used to enter data into the work space are `load`, `fscanf`, `dlmread`, and `input`. In the `load`, `fscanf`, and `dlmread` commands, the data are located in a separate file, usually a .txt file. For the `input` command, the data are entered from the keyboard by the user.

2. Which of the four commands makes the program interactive?

   The program becomes interactive with the `input` command. The user is asked to enter values from the keyboard.

## Review 2.6

1. When there is more than one function plotted on a graph, what are the ways to identify which curve goes with which function?

   Each curve can be given a different color, or a different line type. In each case you can use the legend command to identify which curve goes with which function. You can also use the text command to label each of the curves.

2. What is the name of the function that will allow you to plot several graphs on one page?

   The name of the function that will allow you to plot several graphs on one page is the `subplot` command. The subplot command is not a plot command. It is used to position the several different plots on the page.

Review Answers 247

3. How does one enter Greek symbols into a plot?

   The Greek symbol is entered with \name of symbol.

4. What are the commands that will allow you to enter text onto a plot once the plot has been created?

   In the plot window, click on the Insert option in the menu bar.
   A dropdown menu will appear that contains the following options: X Label, Y Label, Title, TextBox, and others. Click on the item that you wish to enter on the plot. If you select the TextBox option, a crosshair will appear and you can drag it to the location where you wish to start the text, then type in the text that you want to enter into your plot.

## Review 3.1

1. What statement is frequently used to establish two conditional paths?

   The if-else statements.

2. What series of statements is used to establish several conditional paths?

   The if-elseif-else statements.

3. List the various types of logic statements that can be used with the if-else and the if-elseif-else ladder.

   a < b, a > b, a $\leq$ b, a $\geq$ b, a == b, a ~= b.

4. Is the else statement required with the if-else and with the if-elseif-else ladder?

   No.

5. What statement group and a MATLAB's function are alternatives to the if-elseif-else ladder?

   The switch statement and MATLAB's menu function.

## Review 3.2

1. If $y = 3.0 * \mathbf{A}$ and $\mathbf{A}$ is a vector, what can you say about $y$?

   If $\mathbf{A} = [a_1 \; a_2 \; a_3 \ldots a_n]$, then $y = [3 \times a_1 \; 3 \times a_2 \; 3 \times a_3 \ldots 3 \times a_n]$.

2. If $y = 3.0*\sin(x)$ and $x$ is a vector, what can you say about $y$.

   If $x = [x_1 \; x_2 \; x_3 \ldots, x_n]$, then
   $y = [3 \times \sin(x_1) \; 3 \times \sin(x_2) \; 3 \times \sin(x_3) \ldots, 3 \times \sin(x_n)]$

3. If vector **C** = **A** + **B**, what must be true about vectors **A** and **B**.

   Vectors **A** and **B** must be of the same length.

   Each element of **C** will be the addition of the corresponding elements of **A** and **B**.

4. What is the result of the multiplication of two vectors of the same length, say **A** and **B**, and how it should be programmed?

   If $A = [a_1\ a_2\ a_3\ ....\ a_n]$ and $B = [b_1\ b_2\ b_3\ ...\ b_n]$, then the multiplication of A and B is $[a_1 \times b_1\ a_2 \times b_2\ a_3 \times b_3\ ...\ a_n \times b_n]$

   In MATLAB, the multiplication has to be programmed as **A.\* B**.

5. What is the name of MATLAB's function that does interpolation?

   The function name is `interp1`.

6. What are the inputs to MATLAB's interpolation function?

   MATLAB's `interp1` function has three arguments, say (X,Y,Xi), where (X,Y) are a set of known (x, y) data points and Xi is the set of x values at which the set of y values, Yi, are to be determined by interpolation. Arrays X and Y must be of the same length. If Xi is a vector, than Yi will be a vector.

7. What are the outputs from MATLAB's interpolation function?

   In the variables described above the output from MATLAB's `interp1` function are the interpolated values Yi.

## Review 4.1

1. When does it seem appropriate to write a self-written function?

   If you have a complicated program and you wish to break it down into smaller sections, it is appropriate to write a self-written function. Also, if you have a program that requires a series of statements to be repeated several time, it is convenient to place the series of statements in a self-written function. Finally, many MATLAB functions require the user to write a self-written function to describe the problem of interest.

2. In writing a self-written function what must be the first word in the first executable statement?

   The first word in the first executable statement in the function must be `function`.

3. A self-written function usually has both an input and an output. Where does the input come from? Where does the output go to?

   The input comes from the calling program.

   The output from the function goes to the calling program.

# Review Answers

4. If a self-written function has more than one output, how must the output be presented?

   If a function has more than one output, the output must be in brackets.

5. How does a self-written function communicate with the calling program?

   The self-written function only communicates with the calling program through the input and output variables. The exception is when a global statement is contained in both the calling program and the function.

6. What can be said about variables in the self-written function that are not in the input or output arguments of the function and there are no `global` statements?

   If a variable in the function is not in the input or output arguments of the function, then that variable is completely independent of variables in the calling program.

7. Do the variable names in the input and output arguments between the calling program and the function have to be the same?

   No, they only need to be in the same order.

8. If a programmer wishes to write a self-written function, but does not wish to create an additional .m file, what can the programmer do and what is the constraint?

   The programmer can write an *anonymous* function, which is included in the main program and not as a separate *.m* file. The constraint is that it needs to be a single statement.

## Review 5.1

1. Suppose you wish to assign a column vector consisting of string elements, what are the conditions that need to be followed in setting up this column vector?

   The conditions are (a) each string row needs to be enclosed by single quotation marks, (b) each string row must have the same number of columns, and (c) the entire matrix must be enclosed by brackets.

2. Suppose that you had a data file that contains both numerical and text data, what command would you use to read in the data into your main program?

   The command used to read in the data is the `textscan` command.

3. When the command used in reading in the data type described in item 2, what object type does the data go into?

The data goes into a cell array.

4. To assign variable names to items in the object, which of the following three symbols would you use: (), [], {}.

You should use {} symbol.

## Review 6.1

1. What is meant by the term root of function $f(x)$?

   The root of a function is the value of $x$ that makes $f(x) = 0$

2. What is the objective in the search method for determining a root of the equation $f(x) = 0$?

   The objective of the search method is to find small intervals containing the roots.

3. What is the name of the MATLAB function for determining the roots of a transcendental equation of the form $f(x) = 0$?

   The name of MATLAB's function to obtain the roots of a transcendental equation is `fzero`.

4. In MATLAB's function for determining the roots of a transcendental equation, how does one define the function whose roots are to be determined?

   A self-written function should describe the function whose roots are to be obtained. The name of this self-written function should be entered as the first argument in MATLAB's `fzero` function.

5. If you suspect that there is more than one real root, what method should be used in combination with the MATLAB's function to obtain the roots?

   If you suspect that there is more than one root, you should use the search method, in combination with MATLAB's `fzero` function. The search method is used to obtain a small interval in which a root lies and MATLAB's `fzero` function determines the root that lies in that interval.

6. If you are using the search method in combination with the `fzero` function, what can you say about the second argument in the `fzero` function?

   The second argument to be entered in the `fzero` function should be a vector of length 2 specifying the endpoints of the intervals that contain the roots.

   The functional values at the beginning and end of this interval should differ in sign.

# Review Answers

7. **What is the purpose of the global statement?**

   Variables listed in the `global` statement will be common to both the calling program and the called function. Therefore, variables defined in the calling program will be available in the called function, despite the fact that these variables are not input or output arguments in the called function. Of course, the reverse is also true. An item determined in the called function would also become available in the calling program.

8. **If the function f(x) is a polynomial, what MATLAB function should you use to obtain its roots?**

   You should use MATLAB's `roots` function. If the polynomial has complex roots, MATLAB's `roots` function will give the complex roots, whereas MATLAB's `fzero` function will only give the real roots.

## Review 7.1

1. **Given a set of algebraic, linear equations in the form AX = C, where A is the coefficient matrix and X and C are column vectors, what are the two ways for solving for X in MATLAB?**
   
   a. X = inv(A)*C.
   
   b. X = A\C.

2. **Given a large system of algebraic, linear equations of the form AX = C, describe the recommended approach to solving the system of linear equations.**

   First we need to number each equation in the system. We then need to determine the coefficients, $a_{i,j}$, in each equation and the $c_i$, where
   - the $i$ represents the equation number and the $j$ represents the number of the $x$ variable associated with the coefficient.

   Example:

   Suppose we had a system of 10 equations requiring a coefficient matrix, $a(10,10)$ and a $c(10)$ matrix to solve the problem. Suppose the 7th equation was as follows:

   $$-0.6\, x_4 - x_5 + x_8 + 0.6\, x_{10} = 0 \tag{7}$$

   $a(7,1) = a(7,2) = a(7,3) = a(7,6) = a(7,7) = a(7,9) = 0$

   and $a(7,4) = -0.6$, $a(7,5) = -1$, $a(7,8) = 1$, $a(7,10) = 0.6$, and $c(7) = 0$

   After establishing all $a(10,10)$ and $c(10)$ values use MATLAB's `inv` or MATLAB's Gauss-Elimination function to solve the problem, that is, X = inv(A)*C or X = A\C.

## Review 8.1

1. Suppose an experiment produced a set of data and we wished to create an approximating curve, $y_c$, that is a polynomial expression that best fits the data. What is the name of the MATLAB function that will do this?

   The name of MATLAB's function that will do this is `polyfit(x, y, m)`, where $(x, y)$ is the experimental data set and $m$ is the degree of the polynomial. The `polyfit` function returns the coefficients of the polynomial, $a_1, a_2, ..., a_{m+1}$ where

   $$y_c = a_1 x^m + a_2 x^{m-1} + ... a_m x + a_{m+1}$$

2. After executing MATLAB's `polyfit` function, and you wish to obtain values on the approximating curve, $y_c$, at positions $(x_1, x_2, x_3, ... x_n)$ what MATLAB function would you use?

   You would use MATLAB's `polyval(a,X)` function, where vector a is the coefficients of the approximating function and X is the vector of the $x$ position values.

## Review 9.1

1. What is the formula for evaluating the integral, $I = \int_A^B f(x)dx$ by the Simpson's rule?

   First sub-divide the $x$ domain into N equal intervals, where N is an even number giving $x_1, x_2, x_3, ....., x_{N+1}$. Then determine the functional values at the $x$ positions giving $f_1, f_2, f_3, ......., f_{N+1}$. The formula for the integral by Simpson's rule is:

   $$I = \int_{x_1}^{x_{N+1}} f(x)dx \approx \frac{\Delta x}{3}\left[f_1 + 4f_2 + 2f_3 + 4f_4 + 2f_5 + \cdots + 4f_N + f_{N+1}\right]$$

2. What is the name of MATLAB's function for integrating a single variable function?

   MATLAB's function for integrating a single variable function is `integral`.

3. In MATLAB's function for integrating a single variable function how does one define the function to be integrated?

   One needs to write a self-written function that describes the integrand.

*Review Answers* 253

The name of this function should be entered as the first argument in the `integral` function.

4. If the integrand contains nonlinear terms, how must they be treated?

Nonlinear terms need to be entered as element-by-element multiplication or division. Terms involving exponents also need to be treated as an element-by-element operation.

5. Will MATLAB's `integral` function treat improper integrals?

Yes.

## Review 9.2

1. What is the name of MATLAB's function for integrating a two-dimensional function?

    MATLAB's function for integrating a two-dimensional function is `integral2`.

2. List the arguments that go into MATLAB's function for integrating a two-dimensional function.

    The arguments that go into MATLAB's function for integrating a two-dimensional function are

    a. `FUN(X,Y)` which is a function handle for the function that describes the two-dimensional function to be integrated.

    b. `XMIN, XMAX, YMIN, YMAX`, where

    `XMIN≤X≤XMAX` and `YMIN≤Y≤YMAX`

    `XMIN` and `XMAX` are scalars and `YMIN` and `YMAX` may be scalars or function handles.

# Index

**Note:** Page numbers followed by f and t refer to figures and tables, respectively.

## A

Algebraic and transcendental equations, 131–152
   bisection method, 133, 133f
   MATLAB's
      `fzero` function, 134–139
      `roots` function, 139–141
   overview, 131
   search method, 132, 132f
Algebraic, linear equations, 153–167
   gauss-elimination, 161–162
      method, 154–156
   MATLAB's `inv` function, 154
   number of solutions, 162–163
   resistive circuit problem, 159–161
   treatment of large systems, 156–159
Algorithm, 3
ALU (Arithmetic Logic Unit), 3, 19–20
Anonymous functions, 110–112
Application Software, 5
Approximating curve, 178
   with data points, 172f
Approximating function, 169
   and data points, 174f
Arithmetic Logic Unit (ALU), 3, 19–20
Arithmetic operators, 20–21
   in MATLAB, 244
Arithmetic procedure, 3
Arithmetic statement, 19–20
Assignment operator, 19–21

## B

Backslash character (\), 68, 237
Back substitution, 161–162
Bar charts, 63–65, 65f
Bisection method, 132–133, 133f
Bits, 4
Boundary value problems, 205, 227
   finite difference formulas, 228t
   ODE, 227–235

   difference formulas, 227–235, 228t
   overview, 227
   temperature, plot, 231f
`break` command, 85–89
Built-in functions, 7, 19, 133
   MATLAB features, commands, special items, 21–30
   special values, 26–30
   trigonometric and other functions, 21–26
   with vector arguments, 92–93
Bytes, 4

## C

Central Processing Unit (CPU), 3
Characters and strings, 123–128
Colon operator (:), 28–29, 52
Command Window, 10, 18
   MATLAB, 20
   `pi` in, 23
   undock, 23f
Compound logical expressions, 82
Computer programming, 3
Conditional operators, 81–92
   `break` command, 85–89
   `if` command, 81–83
   `if-elseif-else` command, 83–85
   `menu` function, 90–92
   `switch` command, 89–90
Control flow directives, 7
Control Unit, 3
CPU (Central Processing Unit), 3
Cubic splines, 178–182
Current folder toolbar
   and window, 10–11
Curve fitting, 169–185
   cubic splines, 178–182
      MATLAB's, 179–182
   with exponential function, 174–178
   function, MATLAB's, 169–174
   overview, 169

*255*

## D

Data types, 6
`dblquad` function, MATLAB, 194
Desktop/laptop computer system, components, 3–4
Dialog box, 14–15, 14f
    changing folder/path, 16f
`disp` command, 31
`dlmread` command, 52
Dynamic Random Access Memory (DRAM) devices, 3

## E

Editor window, 11, 12f, 13f, 38f
Equivalent triangular set, 161
Exponential, square root, and error functions, 25

## F

`figure` command, 55–57
Finite difference
    formulas
        boundary value problems, 228t
        different end temperatures and losses heat, 228, 228f
    methods, 227
Flash drive, 4
`fminbnd` funtion, 105, 113–114
`fopen` function, 32, 53
`for` loop, 36–43, 88
    syntax, 36
`fprintf` command, 31–32
`fscanf` command, 52–53
FUN argument, 113
Function handle, 110
FUN function, 134, 190, 194
`fzero` function, 133–139
    MATLAB, 145, 149–151
    statement, 136
    syntax, 134

## G

`Gauss-Elimination`, 154, 161–162
    method, 154–156
    program, 162

## H

`global` function, 136
Greek letters and mathematical symbols, 68

## H

Hard drive, 4
`help integral`, command window, 190
Help window, 16f, 17f
`hold on` command, 59–61

## I

`if` command, 81–83
`if-elseif-else` command, 83–85
`if-elseif` ladder illustration, 125
Improper integrals, 190
Initial value problems, 205
    and MATLAB's ODE, 206–214
Input and output arguments, 106
`input` command, 54–55, 107
Input devices, 3
Input, MATLAB, 50–55
    `dlmread` command, 52
    `fscanf` command, 52–53
    `input` command, 54–55
    `load` command, 50–51
Input/output ("I/O")
    commands, 7
`integral2` function, 194–204
    `self-written` integrand function, 194
`integral` function, 190–193
`interp1` function, 93–95
Inverse trigonometric functions, 24–25

## L

`legend` command, 58
Linear interpolation, 85–86, 93
Linear plot, 34–36
`load` command, 50–51
Logical expressions, 82
Loops, MATLAB, 36–50
    `for` loop, 36–43
    `while` loop, 43–50

# Index

## M

Machine language, 5
Main Memory, 3
MATLAB, 9
   arithmetic operators, 244
   backslash operator, 154
   colon operator (:), 28–29
   command window, 20
   computer programming, 1–8
      building blocks, 7
      computer programming, 3
      computer usage, 1–2
      desktop/laptop computer system, components, 3–4
      mathematical model, 2
      methodologies, 6
      needs, 5–6
   curve-fitting function, 169–174
      cubic spline, 179–182
   desktop, 10–13
      PLOTS tab, 34
      windows, 11f, 12f
   features, commands, special items, and built-in functions, 21–30
      special values, 26–30
      trigonometric and other functions, 21–26
   function(s), 26–27
      additional, 139
      built-in, 19, 133
      built-in ode45, 205
      cubic spline, 179
      dblquad, 194
      fminbnd, self-written function and, 105–122
      fzero, 131, 134–139, 145
      Gauss-Elimination, 229, 234
      global, 136
      integral, 190–193
      integral2, 194–204
      interp1, 93–95, 179
      inv, 154, 156, 229, 234
      menu, 90–92
      myfun, 136
      polyfit, 175, 184
      polyval, 170, 184
      pulse(t), 224
      roots, 131, 139–141
      template, 106f
      textscan, 127–130
   fundamentals, 9–80
      assignment operator, 19–21
      input, *see* Input, MATLAB
      loops, 36–50; *see also* Loops, MATLAB
      overview, 9–10
      simple plot commands, 34–36
      variable names and types, 18–19
   graphics, 55–69
      bar charts, 63–65, 65f
      figure command, 55–57
      greek letters and mathematical symbols, 68
      hold on command, 59–61
      interactively annotating plots, 69
      multiple plots, 57–59
      pie charts, 65–68, 66f
      plotyy command, 62
      saving plots, 69
      subplot command, 63
   help window, 17f
   input command, 50
   matrix, preallocation, 29
   menu push button, 82f
   ode45 function, 206, 215, 217
      computer program, 220, 223
      syntax, 206
   ODE, initial value problem, 206–214
   "=" operator, 44
   output, 30–34
      disp command, 31
      fprintf command, 31–32
      printing to file, 32–34, 33f, 34f
   plots, greek letters and special characters in, 237–239, 238t, 239f
   program, 143, 149, 166–167
   programming languages, 6–7
   script, 13–17, 14f, 15f, 237
   version R2016A and R2016B, 54
Matrix algebra, 153–154, 157
Memory cells, 4
Memory/Storage Unit, 3
menu function, 90–92
Moler, Cleve, Dr., 5
mse function, 170

## N

Network interface, 4
Newton–Raphson method, 132
Numerical integration, 187–204
    improper integrals, 190
    MATLAB's
        `integral2` function, 194–204
        `integral` function, 190–193
    of ODE, 205–225
        initial value problem and MATLAB's, 206–214
        overview, 205
    and Simpson's rule, 187–189
Numerical methods, 2

## O

ODE, *see* Ordinary differential equations (ODE)
`ode45` function, 206, 219
`ODEFUN` function, 206
Operating system (OS), 4
Operators, 7
Optical drive, 4
Ordinary differential equations (ODE), 205, 227
    boundary value problems, 227–235
        difference formulas, 227–235, 228t
        overview, 227
        temperature, plot, 231f
    numerical integration, 205–225
        initial value problem and MATLAB's, 206–214
        overview, 205
OS (operating system), 4
Output devices, 4

## P

Pie charts, 65–68, 66f
`plotyy` command, 62
`polyfit` function, 169, 184
    MATLAB's, 175
Polynomial regression, 169
Programming languages, 26
    MATLAB, 6–7
    overview, 5

## Q

`quad` function, 190

## R

Registers, 3
Resistive circuit problem, 159–161
`roots` function, 133, 139–141
    syntax, 139

## S

Scalar and vector operations, 96–99
    addition, 96
    element-by-element operations, 96–98
    scalar times vector multiplication, 96
    two vector addition/subtraction, 96
    functions operations, 98–99
Search method, 132, 132f
Secondary Memory, 4
Self-written function, 105–110
    and MATLAB®'s `fminbnd` function, 105–122
    anonymous functions, 110–112
*Self-written* integrand function, 194
Simple plot commands, 34–36
Simpson's rule, numerical integration and, 187–189, 188f
`size()` command, 154
`spline` function/method, 179
`sprintf` command, 67
String specifiers, 128
`subplot` command, 63
Subscripts and superscripts, 237
`switch` command, 89–90
System software, 5

## T

TeX formatting language, 237
`textscan` function, 127–128
Toolstrip, 10, 13–14
    *open* icon, 33
    *print* command, 32
    save and run icon, 15f

# Index

Transcendental equations, algebraic and, 131–152
 bisection method, 133, 133f
 MATLAB's
  `fzero` function, 134–139
  `roots` function, 139–141
 overview, 131
 search method, 132, 132f
Trigonometric and other functions, 21–26
 complex number, 26
 exponential, square root, and error functions, 25
 functions, 24
 inverse, 24–25
 special values, 23
TSPAN, time interval, 206

## W

`while` loop, 43–50
Workspace window, 11